D0875400

Title page: Goldfinch, *Carduelis carduelis.* Photo by Harry V. Lacey.
Front endpapers: Diamond sparrow, *Zonaeginthus guttata.* Photo by Harry V. Lacey.
Back endpapers: Goldfinch, *Carduelis carduelis.* Photo by Harry V. Lacey.

Photography:
Dr. Herbert R. Axelrod: 155; Helmut Bechtel: 26, 34, 35, 39 (top); M. Davis: 12; Fox Photos: 52; Tom Gardner: 46; Keith Hindwood: 83 (bottom), 86 (bottom), 139 (bottom), 154; Paul Kwast: 38 (bottom), 51 (bottom), 54 (bottom), 55 (top), 62, 66, 70, 78, 87, 90 (bottom), 91 (bottom), 95, 103, 111, 118 (bottom); Harry V. Lacey: 6, 20, 25, 37, 60, 69, 71, 72, 80 (bottom), 99, 102 (bottom), 106 (bottom), 107, 126, 145 (bottom), 146; P. Leysen: 30; Irene and Michael Morcombe: 10, 143; Jurgen Nicolai: 44, 49, 68, 121; Mervin F. Roberts: 14, 15 (bottom), 80 (top); Courtesy of San Diego Zoo: 151 (top); Heinz Schrempp: 11, 94, 130; Louise Van der Meid: 32, 77, 98, 100, 117, 145 (top); Courtesy of Vogelpark Walsrode: 15 (top), 18, 19, 22, 23, 27, 31, 38 (top), 39 (bottom), 42, 43, 47, 50, 51 (top), 54 (top), 55 (bottom), 58, 59, 63, 67, 74, 75, 79, 82, 83 (top), 86 (top), 90 (top), 91 (top), 102 (top), 106 (top), 110 (top), 115, 118 (top), 122, 123, 128, 147, 150, 151 (bottom); Dr. Matthew M. Vriends: 110, 114, 119; John Warham: 17; L. Wilhelm: 127 (top).

Drawings by Hermann Kacher.

Translated by U. ERICH FRIESE.

ISBN 0-87666-841-4

Distributed in the U.S. by T.F.H. Publications, Inc., 211 West Sylvania Avenue, PO Box 427, Neptune, NJ 07753; in England by T.F.H. (Gt. Britain) Ltd., 13 Nutley Lane, Reigate, Surrey; in Canada to the pet trade by Rolf C. Hagen Ltd., 3225 Sartelon Street, Montreal 382, Quebec; in Canada to the book trade by H & L Pet Supplies, Inc., 27 Kingston Crescent, Kitchener, Ontario N28 2T6; in Southeast Asia by Y.W. Ong, 9 Lorong 36 Geylang, Singapore 14; in Australia and the South Pacific by Pet Imports Pty. Ltd., P.O. Box 149, Brookvale 2100, N.S.W. Australia; in South Africa by Valid Agencies, P.O. Box 51901, Randburg 2125 South Africa. Published by T.F.H. Publications, Inc., Ltd., the British Crown Colony of Hong Kong.

BREEDING BIRDS AT HOME

Jürgen Nicolai

Green singing finch hen, *Serinus mozambicus.*

Contents

FOREWORD

Selecting a bird for purchase should be made with great deliberation, because when one acquires a live animal one simply cannot be guided by taste and fancy alone. Whether you buy a bird and add it to your collection depends, of course, totally on the accommodations available, on the present composition of your bird collection and, finally, on whether or not you can meet the special demands that the new bird might have in terms of feeding and care.

The selection of birds available nowadays is larger than ever before; so, the temptation to own a species which is only seldom available or even has been imported for the first time is often overwhelming for cage-bird fanciers. This is understandable and justifiable, when supported by the motivation to discover new and unfamiliar types of behavior and the best possible nutrition for a particular species. Right from the start such an aim is unattainable if the accommodations available are unsuitable, the available space is insufficient and the aviary is already overcrowded or occupied with species which are incompatible with the requirements for the newcomer. Purely collecting "rarities" which are then accommodated in rows of individual cages or in already overcrowded aviaries may satisfy the owner's pride, just as a collection of coasters and matchboxes, but it makes very little sense. If a bird fancier has little room and can devote only limited time to the bird hobby then I advise that he confine himself to one or a few species and study these and their behavior intensively. The most important discoveries which were made on cage and aviary birds during the last decade did not come from "rarities" or "first imports," but instead, without exception, they came from species which are commonly available and of which "everything" should have been known. This, of course, does not imply that rare birds do not provide for interesting and rare observations. Similarly, it does not mean that hobbyists of more common birds are assured right from the start of making exciting and new discoveries. The pure "collector type" whose interest in a particular species rapidly wanes as this bird becomes more commonly kept by other fanciers and the type of conscientious observer who looks for many years with great interest after his birds are so basically different in their attitudes toward keeping birds that the results of their efforts usually also have very different meanings.

Jürgen Nicolai

The red-eared firetail finch, *Zonaeginthus oculatus,* is an Australian grass finch. Unfortunately, captive breeding stocks of this bird are not established.

Opposite:
An Australian grass finch frequently seen in aviaries is the star finch, *Poephila ruficauda.* Star finches exhibit a wide variation in intensity of color—note this brighter race.

Any aviary should contain some branches as natural perches and a variety of nesting facilities.

Breeding in Aviaries

The care of individual birds in cages or in the company of other birds in aviaries where there are no breeding facilities is always somewhat unsatisfactory and should occur only during the transitional period to acclimate the new bird. What we can observe of individually kept birds—that is, feeding behavior, care of plumage and vocalization—is, of course, only a fraction of the entire behavior picture which is exhibited by our birds. The vast inventory of their abilities can only be fully appreciated if we give them an opportunity to reproduce, to build their nest and to raise their young. Consequently, if a hobbyist keeps only individual birds in cages he will never really get to know them; most of their abilities will remain hidden.

There is also another very important point which is to be clearly understood when the decision arises between keeping birds individually in cages or attempting to breed them in aviaries or flight cages. Birds are strongly motivated by instincts and they suffer considerably when they have no opportunity to satisfy these instincts. In many instances there is simply no possibility of affording them the necessary opportunities. How can one possibly provide for a warbler or a robin which is driven by its strong migratory instinct to flutter night after night in its cage, attempting to fulfill the elevating experience of migration flights? Even the largest aviary is inadequate for that purpose, because the bird reaches the next wall within a few seconds after it has started to fly. In these cases we have to simply accept

The most common bird used for fostering is the society finch, *Lonchura striata.* Here a society finch feeds its own young.

Other species that are attentive parents can also be used for fostering. There have been instances where white-hooded nuns, *Lonchura maja* (right), have fostered shafttails. Zebra finches, *Poephila guttata* (below), have been used successfully to foster the young of various species.

the fact that our birds are rather unhappy every night for several weeks and that in their frustrating attempts to break out they exhaust themselves far more than would have been the case in a normal migratory flight.

The situation is not much different during the reproductive period. At that time, even in bird species which are strongly solitary during most of the year such as woodpeckers, chats and thrushes, the drive to win a partner, search for a nesting site and help in raising the young becomes overwhelmingly strong. It is characteristic of most of these innate types of behavior that under heavy pressure, even without a partner to which they are entitled, these behaviors are executed or acted out with a replacement object. Individually-kept shama thrush males fly, at the onset of the breeding season, into a suitable nesting site in caves or similar places to act out their impressive nesting behaviors during which the white crest stands up like a powder puff. Oskar Heinroth reports that a hand-raised reed warbler in spring—that is, at a time when in nature its mates build their skillfully constructed nests among reed grass—would pick up strips of blotting paper ripped loose from the bottom cover of its cage, dip these into the bathing container and carry them around as its nesting material. When summer progresses, even in captive birds, the sex-drive and nesting instinct give way to the motivation to raise young birds. Male song thrushes, robins and blue-throated robins then carry food insects around in their cages for several weeks, apparently in search of young birds to feed them to. No doubt the birds suffer because their strong reproductive and brood-caring instincts are restricted. In many cases, this suffering is certainly stronger than that which they experience because of the restriction of their mobility. Therefore, if the arguments which are advanced by bird fanciers against attacks by those concerned about animal welfare are to be convincing, they should

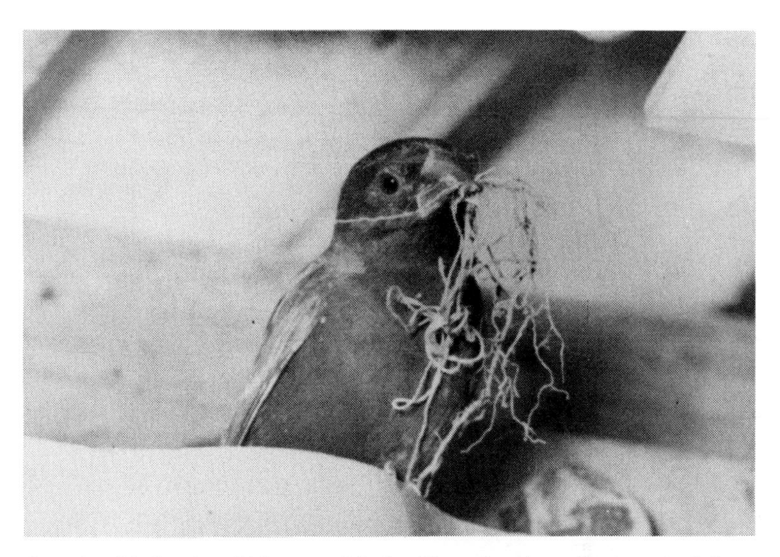

Captive birds should be provided with natural nesting materials or materials which approximate those which the birds would find in the wild. With such materials, the birds will be able to follow their instincts and construct suitable nests.

show, in good conscience, an attempt to fulfill these important requirements of their birds, which are just as important as proper accommodation and a natural diet. It can be said with satisfaction that nowadays more and more bird and animal owners attempt to satisfy these requirements; never before have there been so many bird houses, aviaries and tropical bird facilities built. Moreover, each year many difficult bird species are being bred, whereas a few decades ago even their maintenance in captivity provided insurmountable difficulties.

With this, I have arrived at the third point of my campaign for natural keeping methods. If the bird hobbyist tries to not only keep his birds alive but also affords them the opportunity to reproduce, he is by necessity being educated to a better understanding of the necessary maintenance conditions. If you have ever bred a bird species which you had previously kept only as individual birds in a cage, you know what has to be changed and im-

Both common and rare munias are shown on this and the following page. Not all races of the spice finch, *Lonchura punctulata* (left), are as brightly colored as this specimen. The pectoral finch, *Lonchura pectoralis* (lower left), is a seldom-seen Australian munia. Society finches were developed from wild stocks of the striated munia, *Lonchura striata* (lower right).

Together with the white-hooded nun, the tricolored nun, *Lonchura malacca* (right), and the black-hooded nun, *Lonchura atricapilla* (lower left), are probably the most commonly found munias. The bronze-winged mannikin, *Lonchura cucullata* (lower right), is also often available.

This is an aviary setup used to breed shy birds. Plenty of heather is fastened to the end wall of the shelter, and a heather screen hanging down prevents cats and other small animals from peering into the nests.

proved so that the birds eventually decide to breed. Certain naive suggestions calling for simplification, cost reduction and rationalization in feeding have appeared for decades with great regularity in the bird hobbyist literature. They have reached their apex with the invention of artificial ant pupae and in the proposal of dietary yeasts and universal poultry grain as ideal foods for insect-eating birds. These suggestions are invariably made by beginners or by keepers who keep individual birds and who realize only at the death of their birds the feeding mistakes they have made. Whoever wishes to breed birds cannot accept such gimmicks, because he will learn from his birds, quickly and unmistakably, that they do not play along. Only a few domesticated and partly domesticated species can accept unfavorable conditions, and major errors are here immediately detectable in the vitality of the juvenile birds.

The normal progression of reproductive behavior depends on a number of factors. Prerequisite is, of course, that both sexes are perfectly healthy and that they feel safe. In overcrowded aviaries where some birds are constantly pursued by stronger and more aggressive birds, or in cages which are in close proximity to human traffic or on which cats climb around, only a few bird species—such as canaries, budgerigars, zebra finches and a few other domesticated species which have become immune to disturbances—should be bred. Newly imported birds from the tropics are invariably so disturbed by the enormous change incurred during capture and transport, at the bird dealer and through the dramatic change of the length of day that they themselves will not make any breeding attempts, even if they have been caught at the onset of the breeding season and even if they are given favorable conditions in captivity. As a rule, these birds immediately begin to molt and are then able to reproduce during the next spring. Only after the first molt in captivity can these birds be considered to be properly acclimated.

Apart from these physiological prerequisites, other environmental conditions have to be fulfilled if a bird is to be brought into reproductive condition. Every European field ornithologist knows that a reed warbler can only be found in large stands of reed grass, a skylark only in open meadow areas or fields, and a forest warbler only in a beech tree forest. Only in a species-correct habitat can free-living birds be expected to exhibit a normal development of their breeding behavior, because only there do they find suitable nesting sites and, later on, sufficient food for the progeny. Of course, one can only duplicate to a limited degree the natural environment of a bird species, even in a large aviary. This is not absolutely necessary, because birds can accept a number of limitations if, at least, some of the major characteristics of the habitat and particularly a species-correct nesting site are available. Inhabitants of dense reed

Aside from the magpie mannikin, *Lonchura fringilloides* (left), which has lately been imported with some frequency, the munias on this and the following page are seldom available. The black and white mannikins, *Lonchura bicolor* (lower left) and *Lonchura bicolor nigriceps* (lower right), are both smaller than the magpie mannikin.

Yellow-rumped munia, *Lonchura flaviprymna* (right). Javan munia, *Lonchura ferruginosa* (lower left). Bib finch, *Lonchura nana* (lower right).

patches, such as the reed warbler, are not particularly happy in a barren aviary which has only a few horizontal perches. Of course, one cannot possibly expect that they are going to make any breeding attempts there. Birds of the open plain are equally unsatisfactorily accommodated in a densely planted flight area. However, the more commonly kept aviary birds are not inhabitants of such specialized habitats, but live in forests, in bush land and in the savannah. For these an aviary planted with bushes and small trees, but with ample open space, is quite suitable.

The simplest method is, of course, to provide nesting opportunities for cave (i.e., nestbox) brooders. With the exception of woodpeckers and a few other birds which make their own breeding cavities (it must conform to a specific depth and have an entrance hole with a special diameter), cave breeders in general have to accept in nature highly variable tree and branch caves. These species are, therefore, completely satisfied with natural caves or nest boxes which in depth and entrance hole diameter conform only roughly to the bird's size.

Most of the estrildid finch species accept artificial nest boxes. In these they prefer to build a nest just as they would in the wild. Therefore, these baskets or nest boxes should not be too narrow. Unfortunately, most commercially available finch baskets, which are only the size of a man's fist, are too small for nearly all species. These birds should have a nest box which is about 14 cm long, 11 cm wide and 12 cm high, with a removable lid. One of the smaller sides should remain either half open or have a U-shaped entrance hole. Half-opened nest boxes are preferred over enclosed boxes by orange-winged pytilias, melba finches, aurora finches, striped waxbills, twin-spots, and most of the fire finches. Enclosed nest boxes are more suitable for parrot finches, especially the blue-faced, red-headed, and Peale's parrot finches. Gouldian finches require larger nest boxes, such as budgerigar nest boxes of the elongated format. Par-

European goldfinch with fledglings and bullfinch juveniles. Both of these birds can be kept most successfully in an outdoor aviary, and a variety of foods will help to maintain the colors of both species.

ticularly large but not enclosed nesting facilities are required by the diamond sparrow. It builds a large nest the size of a soccer ball, sometimes even larger. It is advisable to offer to this species small (30 cm long, 20 cm high and 20 cm wide) wooden cages with wooden bars and a solid bottom. There the birds can build their large nest made of bulky material.

Some grass finches will accept artificial nest boxes only in exceptional circumstances. These include the purple grenadier, the violet-eared waxbill and the black-bellied fire finch. The first two species prefer to build their nests in dense trees or bushes; the black-bellied fire finch likes to build between or under dense tufts of grass in low bushes overgrown by grass, between rocks or dried branches and, only under exceptional circumstances, in baskets attached low above the ground. One can provide additional nesting

Silverbills are among the more quietly colored munias. Indian silverbill, *Lonchura malabarica* (left). Pearl-headed silverbill, *Lonchura caniceps*. African silverbill, *Lonchura cantans*.

The grey singing finch, *Serinus leucopygius* (below) is reputed to be one of the best songsters among finches, but it is also aggressive toward birds of similar size. The quail finch, *Ortygospiza atricollis muelleri* (right), spends a lot of time on the ground and so would best be housed in a planted aviary.

facilities for these species by introducing dense bundles of pine branches or similar material at different heights in the aviary.

Open brooders such as true finches often accept the wire nests equipped with a reed basket used by canary breeders. Cover the wire sides with cloth or artificial pine branches and thus give the birds a feeling of security. Nest blocks made from a branch 8 cm to 12 cm thick and 10 cm long with a depression carved into the open cut area are also accepted. Around the circumference of this block one drills thin holes obliquely toward the outside, into which one inserts stiff wire with attached artificial pine twigs. These nest blocks were utilized to breed the linnet, the common redpoll, the serin, the citril finch, the Eurasian siskin, the goldfinch, the Chinese greenfinch *(Chloris sinica),* and the bullfinch. Even waxbills such as gold-breasted waxbills and fire finches use these as support for their spherical nest structures. All these nesting facilities are attached to the wall of the indoor or outdoor aviary and not among branches where the breeding birds would be disturbed by the non-breeding inhabitants of the aviary.

Even among true finches there are species which will sometimes accept such artificial nesting facilities. These include especially the lesser goldfinch, the black-headed siskin, and Yarrell's siskin. Breeding these species is usually only successful when these birds live in a densely planted outdoor aviary which includes a two-meter tall juniper tree or something similar where they build their nest made out of coconut fibers, grasses and other plant materials among the branches.

Most pigeon and dove species are not particular as far as their nest site selection is concerned. They will accept a variety of small, low boxes and baskets and will build there a simple nest made of thin twigs and short coconut fibers.

Among the common insectivorous birds, only the shama thrush is a cave or semi-cave brooder. Initially, females carry a large amount of dried leaves, moss or paper strips into the large nest box, and there they build a nest out of 10 cm to 12 cm cut coconut fibers. The nest depression is often padded with black horse hair. Semi-cave brooders also include the rock thrush and the blue rock thrush. They build their nests in crevices, rock piles, between boulders or in the walls of deteriorating buildings. The rock thrush likes to build its nest on top of beams which are attached directly below the roof of the indoor aviary.

Warblers are known to be open brooders and can be easily persuaded to breed in captivity when the aviary is planted with dense bushes (dogwood, raspberry bushes, small pine trees, juniper bushes and brambles). The blackcap has a preference for abandoned warbler nests, which can be attached at suitable places in the aviary. Siberian rubythroats bred for the first time in England in 1962 in half-open boxes 20 cm square. During that year the same breeder also succeeded with a brood of Himalayan rubythroats. This particular pair had built its nest immediately above the ground in a grass-overgrown hazelnut bush.

Nightingales and thrush nightingales, when kept in aviaries, like to build their nests among branches close to the bottom of such bushes as hazelnut and spirea and among the fresh outgrowth of recently chopped-down willow trees and alder trees which are surrounded by a dense growth of grass or nettles.

The flycatchers usually are semi-cave brooders. In nature, they build their nests in crevices and excavations, rotting logs (little blue-and-white and verditer flycatchers), in burrows along steep slopes (verditer flycatcher), among tree roots or in caves in moss-overgrown rocky cliffs (rufous niltava), but in an aviary situation they will accept a variety of artificial nesting sites. Each bird species has an innate knowledge of the species' typical nesting site; also, individual species know exactly which material is suitable for

The chestnut-breasted mannikin, *Lonchura castaneothorex,* is an Australian grass finch well established in captivity.

Opposite:
Although subduedly colored, the cherry finch, *Poephila modesta* (above), is an attractive little bird. Note the bill color on the shafttails: *Poephila acuticauda* (left) and *Poephila acuticauda heckii* (lower right). The shafttail with the brighter bill color *(P. a. heckii)* is more commonly found in the United States.

Colonially inclined finches will use the apartment-like nest shown here (left). Note the tough-foliated bushes used in this planted aviary (below).

the construction of their nest. Here, too, most bird species are willing to compromise, if at least the most important characteristics of the nesting materials are available. This can also be seen in a number of free-living birds. Ringneck doves often build their nests wholly or in part from pieces of wire which they pick up from some industrial area. These pieces of wire are usually of the same thickness as the dried branches which these birds would normally use. When collecting nest material, the male selects the various parts with its beak, according to just a few criteria—the material must have a certain length, rigidity and strength; thus the bird does not differentiate between twigs and pieces of wire of identical thickness. This often leads to the construction of solid wire nests in which, unless the eggs are punctured by a protruding piece of wire, brooding runs its normal course.

In an aviary we have ample opportunity to convince ourselves that our birds do not insist upon using the same material which they used in the wild to build their nest. It is only important that the material has a consistency which permits the birds to execute their inherited nest construction behavior. Only this guarantees the construction of a typical nest. Thus, shama thrushes which build a nest out of dried leaves accept without hesitation small strips of paper for the same purpose. In fact, free-flying shamas sometimes completely omit the dried leaves if they find a supply of usable paper strips. Oriental white-eyes *(Zosterops palpebrosa)* use cotton wool or similar material which they initially gather in large quantities to wrap around branches so that a fist-size pile is formed with a depression on its middle which is then lined with more solid material. In nature these birds probably use some plant fibers, and in an aviary they usually accept coconut fibers. An Oriental white-eye pair ready to breed often looks in vain for suitable material among that which is customarily provided in small bird aviaries. Neither grass nor feathers fulfill the re-

Because of their beauty, New World buntings are very popular cagebirds in Europe. Indigo bunting, *Passerina cyanea* (left); rainbow bunting, *Passerina leclancheri* (lower left); painted bunting, *Passerina ciris* (lower right).

These Asian birds are insectivores: orange-headed ground thrush, *Zoothera citrina* (left); blue-headed rock thrush, *Monticola cinclorhynchus* (lower left); rufous-bellied niltava, *Niltava sundara* (lower right).

quirements, and even coconut fibers, which are only used in the advanced state of nest construction, are unsuitable. These birds are now being offered short cotton threads, such as those used by canary breeders for their birds, and they get greatly excited and carry large bundles of these to their selected nesting site. However, now it becomes obvious that this material, in its appearance as well as in its softness and white color, does not conform in its other characteristics to the required cotton wool. It can be wrapped around branches, but the birds are unable to pull at it with their beaks and feet to safely secure it. All their attempts fail and finally all that is left of the selected nesting site are a few threads hanging from the branches. If we provide cotton wool or raw cotton in the aviary, the birds finally have what they need. Within a few hours a sizable amount of this material is taken to their nesting site among the branches, wrapped around them and interwoven with coconut fibers to secure the nest. On the following day the birds will start on the nest depression, forming it with coconut fibers into a solid and deep container, and a few days later the first uniformly colored blue egg is in the nest.

By the way, the building material of nearly universal utility is coconut fibers. They replace a variety of materials which the birds normally use in the wild. For various grass finches this material replaces reeds; for the bullfinch it replaces finely branched roots; for the shama thrush it is material to line its nest; and for the small dove species these fibers replace the grass stems and finely branched roots which they use for the construction of their lightweight nests. That these pliable and easily utilized threads meet the requirements of the birds is proved by the following observations. Free-flying bullfinches, which I released at the onset of the breeding season, lived and bred together with wild bullfinches in a typical bullfinch habitat. Invariably, they built their nests in small evergreen trees, and for the nests they almost exclusively used coconut fibers

This bullfinch is sitting on five eggs laid in a nest constructed with coconut fibers.

which they collected near the aviaries. Even our shama thrushes, which were enjoying unlimited freedom, often looking for food several hundred meters away from our house in the forest, built their nests of strips of paper and coconut fibers which they found in the room where their nest cave was located. In order to find out which materials these birds would use in nature, I once removed all available coconut fibers. This was followed by a complete stoppage of all nest-building activities for a short period of time; the female looked high and wide and finally came back with a bundle of coconut fibers which it must have found in the garbage dump.

Of the twenty-seven species of rarely imported South American birds called seedeaters, the lined finch, *Sporophila lineola* (left), is one of the more attractive. The black-crested finch, *Lophospingus pusillus* (below), is a frequently available bird from South America. Efforts to establish a captive breeding population should be made.

The red-crested cardinal, *Paroaria coronata* (right), and the crimson pileated finch, *Coryphospingus cristatus* (below), are South American birds which breed readily when the proper conditions are available.

Other very popular nesting materials for many grass finches include dried stems and leaves of various grasses. Parrot finches prefer broad-leaved grasses. In order to make the largest possible selection of various nesting materials available, we also provide moss, cotton strands (for serins and canaries), sisal hemp from ropes cut into 10 cm sections, wool, raw cotton and goat and dog hairs. The larger dove and pigeon species often like to use rice roots to construct rather magnificent nests. Totally unsuitable are agave threads and oakum, which will easily wrap around the legs of ground birds such as quails and partridges, and thus endanger the birds. Coconut fibers and rice roots are available in bundles from brush and broom factories. Do not purchase threads in a ball, as they are occasionally found in seed stores or pet shops. These are useless and even dangerous for birds when they are building nests because this material contains loops and knots in which a bird can become entangled, break a leg or even strangle itself.

It is a long road from building the nest to laying the eggs and brooding these until the young nestlings are finally ready to leave the nest. During these weeks mistakes made by the birds and the keeper can endanger the desired breeding success. A keeper who spends much time with his birds can enter the aviaries without causing a wild panic among the birds. Only those birds closest to the person entering the aviary will move away, but this does not have the character of a panic flight; it is more like yielding a "right of way." The larger the aviary and the smaller the number of birds in it, the more trusting are the birds. Every bird keeper should do his best to maintain this trust and even deepen it if possible, because it enables easier observation during the courtship and during the brooding period and when the young birds are being looked after. It also permits any necessary access to the nest in order to save an endangered brood. If, however, nets are being used in the

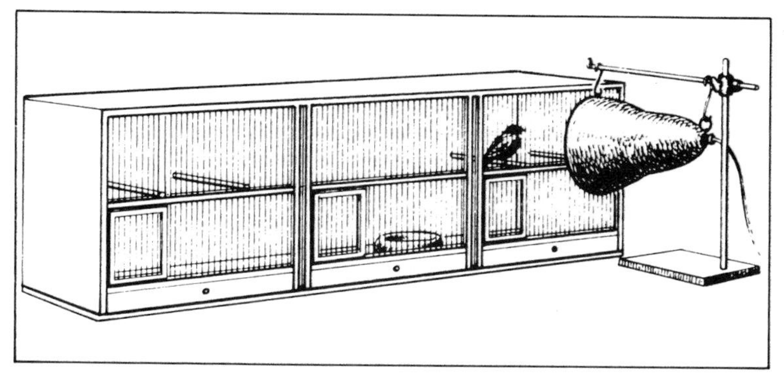

This hospital cage is set up so that the bird can either remain close to the heat lamp or move away, as it chooses.

aviary, the birds will never become tame because the use of such nets represents a terrifying experience which resembles that of being pursued by a bird of prey.

If a bird has to be caught during the day, perhaps because it has become sick or incompatible and attacks other birds, there is only one method of catching it without disturbing the other aviary inhabitants: with a squirt gun filled with lukewarm water that particular bird is wetted down—possibly from a distance through the wire—until it is so wet that its flight is impeded. Then one can catch the soaked bird relatively easily because any flight attempt invariably ends quickly on the bottom of the aviary. Such artificial "rain" neither hurts the bird to be caught nor affects the other birds in the aviary. Of course, a soaked bird will have to be immediately transferred to a hospital cage with an infra-red bulb and has to remain there until it is completely dry again. Incidentally, such a squirt gun is also used for recapturing escaped birds, provided one can get close enough to the escapee. However, in such cases the escaped bird must receive the full force of a shower, because if the bird gets upset and flies away one cannot usually get close enough again to repeat this procedure. In less urgent cases one can hold off with the capturing until it is dark. Then it is usually very easy to catch the bird with the aid of a flashlight.

To prevent a wild population from becoming established, recent legislation has prohibited importation of the Java sparrow, *Padda oryzivora* (left), into the United States. Like the Java sparrow, the red-headed finch, *Amadina erythrocephala* (lower left), and the cut-throat finch, *Amadina fasciata* (lower right), are reliable breeders.

The popularity of European finches as cagebirds is exemplified by the number of hybrids which have been produced—a European goldfinch and bullfinch hybrid (right). The greenfinch, *Carduelis chloris* (below), is a popular candidate for crossbreeding because it breeds readily.

A female lesser goldfinch, *Carduelis psaltria*, brooding.

If there is a fair amount of trust and confidence between the keeper and his aviary birds, nest inspections can be made without any hesitation. There is a fair amount of conflicting opinion among bird fanciers whether such inspections should or should not be made. I know many hobbyists who reject this in principle and instead prefer to be surprised as to whether or not the brooding and the raising of the young progress normally and the young birds eventually leave the nest. At some time or other most of them have had the experience of having some pairs not returned to their nest after such an inspection, and the brood was subsequently lost. However, no bird species can afford to abandon the brood at the first slight disturbance. Birds in the wild are often disturbed by people, grazing cattle, deer and larger birds which sometimes pass the nest. If they would always abandon the brood over such minor disturbances, the majority of bird species would be extinct by now. Also, the "deception behavior," which consists of a brooding bird pretending to be sick or injured and thus guiding the intruder away from the clutch, of course only makes sense if the brooding bird intends to return to its nest after the disturbance. Only if the brooding birds leave their nest in total panic and under the impression that their own lives are immediately threatened do they, in many cases, not return to their nests. A good case in point would be if one removes a brooding female during the night from its nest in order to attach a band. Especially during the onset of the brooding season, many birds will then abandon the clutch, and in these particular cases such a reaction is well in the spirit of the preservation of the species.

Thus, if aviary birds are not frightened to death during a nest inspection and if they do not leave their nest in total panic, which can happen if the aviary is too small, such a disturbance should be quite tolerable. Of course, there are always exceptions to the rule. For instance, Peter's twin-spot *(Hypargos niveoguttatus)* takes rather unkindly to a nest inspection, especially at the onset of brooding. Although one does not get the impression that the birds leave their nest in panic, both partners emit warning sounds, often for up to half an hour, and quite frequently they will give up the clutch. So far we do not know why this particular species is so touchy in this respect.

Frequent nest controls have many advantages, and many a brood can be saved which would have been lost without the aid of the bird keeper. However, during brooding there is little one can do, other than possibly discovering a slightly damaged (cracked or indented) egg. Such eggs can be saved by placing a drop of all-purpose glue or nail polish onto the crack and then waiting until the adhesive agent has completely dried. Of course, such emergency measures are only successful if one discovers the damage in time, before the embryo dries out and dies. I have several birds in my

The Sydney waxbill, *Aegintha temporalis,* is a bird that has almost disappeared from captivity because breeding stocks were not managed properly.

Opposite:
Gouldian finches, *Chloebia gouldiae.* The white-breasted mutation (upper right) can occur with any of the natural face colors: yellow, black, or red.

aviaries which came from such "glued" eggs, and they would have died and dried out as embryos had I not applied this particular measure.

More important are nest inspections after the nestlings have hatched. We have to see daily whether their crops are filled with grain. Even in insectivorous birds one can see whether or not the nestlings are being well fed. As long as their little stomachs are solidly filled, their digestive tract does not protrude and respiration is regular, there is no need for any concern. Newly hatched nestlings or those which are only a few days old but are obviously not being fed can usually be saved if one has an opportunity to add these to the nest of a pair of the same species which has young of the same age. Alternatively, especially with grass finches, the eggs can be given to a pair of brooding society finches which have been sitting on their eggs for some time. There is really no objection to taking occasional advantage of society (Bengalese) finches as foster parents. The young birds will thrive—especially when their dietary needs are not that much different from young Bengalese—just as well as with their own parents, and they grow up to be quite normal and healthy birds. A prerequisite is, of course, that the foster parents have available to them all necessary dietary items which are needed to raise their adoptive young. These items should include ant pupae, egg mixture, germinated spray millet, chickweed and half-ripened grass seeds. Only species which require extensive animal food (such as melba finches) would be inadequately fed by Bengalese foster parents, which change over to a grain diet within a few days. It is categorically wrong to say the grass finches raised by Bengalese finches are invariably useless for further breeding because they build incomplete nests or no nests at all, they do not sit on the eggs or they do not raise their young, as is often claimed. Problems arise only if one uses society finches as "brood machines" and misuses valuable grass-finch species by causing them to lay eggs too

A yellow-rumped serin hen, *Serinus atrogularis,* broods a seventeen-day-old chick.

frequently. This then causes not only premature exhaustion of the breeding birds but also gives rise to the genetic development of negative characteristics. Indeed, the latter may lead, after many generations, to a deterioration of brood-caring behavior. At this stage I do not want to elaborate on the problem of "imprinting" upon the foster parent species. It should be sufficient to say that the reader must be aware that imprinting upon the foster parents occurs, but this is not a deciding obstacle for later pair bonding with other birds of the same species.

When larger nestlings which have survived the first ten critical days and have developed normally are suddenly neglected, this is invariably caused by a particular environmental condition affecting the parents. Some birds, particularly grass finches, often prematurely start the reproductive cycle again by preparing for a new brood. A regular consequence of this is that the nestlings which at

Three of the popular *Amandava* waxbills: green avadavat, *A. formosa* (left); gold-breasted waxbill, *A. subflava* (below left); strawberry finch, *A. amandava* (below right).

Though the orange-cheeked waxbill, *Estrilda melpoda* (below), along with the red-eared waxbill, is consistently available, it has not become established in American aviaries. Enough black-cheeked waxbills, *Estrilda erythronotus* (right), have been imported into the United States lately to establish this species.

A rare photo of a fledgling being fed by a parent still in flight.

that age are not being guarded by the parents during the day are being fed less and less until all parental attention ceases. The parent birds now meet each other outside the nest, and since they are not both occupied in their search for food for the young, they endanger the development of previous nestlings. If one makes a regular nest inspection, the broods in such cases can usually be saved. After observing the birds for several hours, it is easy to determine which one of the two partners still seems interested in the nestlings and which one still feeds them occasionally. Usually the male is the instigator for a new brood. He follows the female with courting attempts, trying to entice her to a newly selected nesting site, leaving the female little time to look after the young. If this has been determined to be the case, the male should be removed from the aviary and taken to another room, one far enough away so that he can not remain in vocal contact with the female. After the female has been looking for and calling for the male for some time, she will often resume feeding the nestlings, often as early as within about half an hour.

If neither of the two partners shows up at the nest anymore and if time has become precious, both of them must be removed. We take these two birds together with their young and the nest and place them into a larger cage (at least 50 cm long) which is then placed into a quiet room. The cage should be directly at a window, ideally on the windowsill, and should be covered for the first few days with a thin white cloth or gauze. This will keep all disturbances away from the birds. This totally changed environment will extinguish their premature breeding attempts, and they will usually resume feeding their own nestlings.

Of course, much depends upon how early we detect the parental neglect and whether we immediately initiate corrective measures. Invariably, the parents will not sit on the nest anymore at the new location, and thus the nestlings have to be large enough and have at least some plumage

The photos on this and the following page allow the opportunity to compare three similar estrildids: St. Helena waxbill,
(left); rosy-rumped waxbill, *E. rhodopyga* (below); red-eared waxbill, *E. troglodytes* (opposite above). The differently colored black-crowned waxbill, *E. nonnula* (opposite below), is also an estrildid.

when the entire family is being transported to new quarters. From that point on, the nestlings have to do without the warmth of the attending parent bird. If the nestlings are still too small, such a transfer of the nest is, of course, out of the question.

When the young birds are neglected after they have become fledged, the only thing to do is to remove both parents. Each partner is then accommodated with some of their youngsters in a new cage which is totally separated and out of sight from the other partner and from the previously occupied aviary, so that there is no vocal contact. In most grass finches it is only the male which feeds the fledglings, so it is sometimes possible—after an observation period—to place all fledglings together with the male. However, it is better if we can motivate both parents to continue feeding the young birds. Thus, we maintain the possibility of an exchange of youngsters later on, in the event that one of the two parents does not maintain the support of the young birds.

Finally, a word of warning which will hopefully be heeded by the readers of this book. During the last few years I have visited many bird fanciers and have had a look at their facilities and stock. In doing that, I have seen in many instances excellent facilities with correspondingly good breeding successes and I have seen young birds of species which a few years ago were still considered to be extremely difficult to raise or even to breed. When I visited the same breeders again a few months later, invariably I found very few of those recently bred juvenile birds still there. Most of them had been sold to other bird hobbyists who had the necessary cash on hand. In a few instances I made an effort to follow the fate of the birds that had been sold, and invariably I learned that the majority of these had been placed in circumstances where they could not have a chance to reproduce. Usually the ornithological knowledge of the new owner and the accommodations were limited to such a degree that it was insufficient even for the hardiest species. It is probably not overly exaggerated when I say that 80% of these particularly demanding and very valuable young birds did not live beyond the year of their birth.

It seems to me that in many bird breeders the feeling of responsibility for the continued survival of the birds they have just bred ceases the moment these birds are on their way to the first available customer. It is this fatal attitude which is the reason that with the exception of the red-headed parrot finch, not one of the non-domesticated parrot finches survived the years of inadequate food supply during the last world war. For instance, today—only a few years after the Australian government banned all exports of its animals—the Sydney waxbill *(Aegintha temporalis)* has already nearly disappeared from the aviaries of grass-finch fanciers. Several bird species which are now thought to be extinct in captivity could have been preserved by concentrating the stocks and through planned breeding by a few selected breeders.

When you have finally succeeded in breeding a difficult species, after great difficulty and many significant sacrifices in terms of time and effort, the young birds should not be disposed of to the first "bird fancier" who comes into your house and offers the necessary cash. All this effort only makes sense when you go beyond the stage of looking for medals, certificates or other show prizes and you assure the survival of your birds and ultimately the survival of the species in captivity. In order to accomplish this you have to make sure that the accommodations, experience and abilities of the new owner are familiar to you before you sell your birds to him. If you have to decide the fate of your bred birds—that is, birds which you cannot keep any longer—you should act as if the parent pair and its progeny were the last representatives of the species and as if you had the sole responsibility whether this species becomes extinct or survives through planned, continued breeding.

The yellow-bellied waxbills, *Estrilda melanotis quartinia* (left) and *E. m. bocagei* (male below left, female below right), are seldom imported.

Also infrequently available are the dark fire finch, *Lagonosticta rubricata* (right), and the black-faced fire finch, *L. larvata larvata* (male below left, female below right).

Left to right: Senegal combassou male, cordon bleu female, red-billed fire finch male and cordon bleu male.

Grass Finches and Waxbills

VINACEOUS FIRE FINCH, Lagonosticta larvata vinacea

Range: Western Africa.

Habitat: Bamboo stands in the proximity of rivers.

Usually juvenile birds are imported which are no older than six to eight weeks. However, since these are extremely delicate, they have to be acclimated in high heat in a quarantine cage with an infra-red lamp. The birds suffer invariably from infectious intestinal diseases; therefore, strict isolation (four to six weeks) especially from all other fire finches, Terramycin treatment and adequate insect food are a must. The birds are properly acclimated only after adult plumage is obtained. Diet: Small grain millet, canary seeds, egg mixture, ant pupae, small wax moth caterpillars and spiders. Well-acclimated birds will breed without difficulty in protected, well-planted aviaries.

DARK FIRE FINCH, *Lagonosticta rubricata*

Range: Most of tropical Africa.

Habitat: Among bushes in tall grass of savannah, along the edges of forests and among bushes in open forests.

Newly imported birds are very delicate and must be acclimated like the previous species. Males and unpaired females produce highly variable vocalizations with pleasant sounding, flute-like calls in variable pitches. They breed reliably and usually raise broods without difficulty, particularly when fresh ant pupae are offered.

Bar-breasted fire finch, *Lagonosticta rufopicta.*

Opposite:
Lavender finch, *Lagonosticta caeruiescens* (above). Jameson's fire finch, *L. rhodopareia jamesonii* (below left). Black-tailed lavender finch, *L. perreini* (below right).

JAMESON'S FIRE FINCH, *Lagonosticta rhodopareia*

Range: Eastern and southern Africa, from Eritrea to Angola.

Habitat: Bushes along rivers in the dried thorn bush areas and in open forests.

Maintenance and feeding are the same as in the closely related dark fire finch. Hardy vocalization resembles that of previous species, but it is shorter, is presented at a slower speed and is characterized by a prolonged canary-like trill. It has been bred several times.

BLACK-BELLIED FIRE FINCH, *Lagonosticta rara*

Range: Sierra Leone and Nigeria to western Kenya.

Habitat: In bushy, tall grass of the savannah, in open acacia forests and among rubble of high plateaus.

Usually imported together with the vinaceous fire finch, predominantly as juvenile birds. Equally delicate and often carries contagious intestinal infections. Careful acclimation in a quarantine cage with infra-red lamps and Terramycin treatment is required. The birds will immediately take small wax moth caterpillars and later accept ant pupae and egg mixture. Feeding: Same as the vinaceous fire finch. Breeding is difficult.

BAR-BREASTED FIRE FINCH, *Lagonosticta rufopicta*

Range: From Gambia (West Africa) to the southern Sudan.

Habitat: Among tall grass and bushland of the savannah, especially along river banks and close to human habitation.

Newly imported birds are just as delicate as the vinaceous fire finch and black-bellied fire finch. Very shy, frightens easily and, therefore, is unsuitable for cage maintenance. During the breeding season it is extremely aggressive toward other fire finches. Although susceptible to health problems during wet, cold weather, it has been bred repeatedly.

BROWN FIRE FINCH, *Lagonosticta nitidula*

Range: Angola to southern Congo and northern Rhodesia.

Habitat: Swampy areas with low-lying bushes and reed stands, occasionally also in higher elevations (up to 1,800 m).

Much hardier than the bar-breasted fire finch. During the breeding season it is aggressive toward other fire finches; therefore, the species should not be kept together with other fire finches. Has been bred repeatedly.

RED-BILLED FIRE FINCH, *Lagonosticta senegala*

Range: From Western Africa (Senegal) to Eritrea, southward to the Cape.

Habitat: In dried areas close to open water. Commonly around human habitation, sometimes even entering houses. Fairly tame.

The Western African race is frequently imported and is one of the more commonly available grass finches. This species is less difficult than the other fire finches, but these birds are susceptible to wet, cold climate, and they do not live very long. The Eastern African race is much hardier and more durable. It breeds easily and raises the young reliably. Feeding: Same as for the vinaceous fire finch.

VIOLET-EARED WAXBILL, *Uraeginthus granatina*

Range: Southern Angola to Rhodesia and southern Mozambique.

Habitat: The dry thorn-bush steppe.

Between the twenty-fourth and thirty-fifth day plain-colored, yellow-brown juveniles develop the violet cheek spots and the blue forehead band of the adults. From that point on the sexes can be clearly distinguished (males with dark violet cheeks, females with pale violet cheeks). The remainder of the juvenile plumage will be lost at the age of about two to three months. Newly imported birds are very

The most common fire finch is the red-billed fire finch, *Lagonostic-ta senegala.*

Opposite;
Three *Uraeginthus* species: violet-eared waxbill, *U. granatina* (above); purple grenadier, *U. ianthinogaster* (above right); cordon-bleu, *U. bengalus* (below).

Three-week-old gold-breasted waxbill, *Amandava subflava,* just prior to leaving the nest.

delicate and have to be carefully acclimated at elevated temperatures and with sufficient insect food (wax moth larvae, freshly molted meal worms, ant pupae, and white worms). Particularly susceptible to wet, cold climates. Very aggressive toward birds of the same sex as well as toward the closely related purple grenadier and also quite often towards cordon bleus. Therefore, should be kept only in pairs when in the company of other waxbills. Feeding: Small grained millet, shelled millet, silver millet, egg mixture, ant pupae, insects, white worms, half-ripened grass seeds and chickweed.

PURPLE GRENADIER, *Uraeginthus ianthinogaster*

Range: Eastern Africa from southern Ethiopia and Somalia to southern Tanzania.

Habitat: The dry thorn-bush steppe.

Less delicate than the violet-eared waxbill, but equally

aggressive toward specimens of the same sex and toward the closely related violet-eared waxbill and also, sometimes, towards cordon bleus.
Feeding: same as violet-eared waxbill.

ORANGE-WINGED PYTILIA, *Pytilia afra*

Range: Eastern Africa from southern Ethiopia toward central Mozambique and Angola.

Habitat: Bush-covered grasslands, along the edge of forests and in open forests.

The orange-winged pytilia often arrives in better condition in pet shops than does the melba finch, since it is less delicate during transport and at the dealer because it eats millet. In the long run, however, these birds can only be maintained with sufficient animal food. Breeds easily and raises the young without difficulty when fresh ant pupae are available. Feeds on grass seeds and insects.

Two juvenile male yellow-bellied waxbills, *Estrilda melanotis bocagei.*

A male blue-capped waxbill, *Uraeginthus cyanocephala.* Females of the three blue *Uraeginthus* species can be distinguished by variations in bill color and by the extent of blue plumage.

Although the blue is often more extensive on the males, the sexes of the blue-breasted waxbill, *Uraeginthus angolensis*, are best distinguished behaviorally.

A pair of melba finches, *Pytilia melba*. The melba on the left is a male.

MELBA FINCH, *Pytilia melba*

Range and habitat: Open savannah areas of Africa from Senegal to Eritrea and from there southward throughout the eastern Africa to Mozambique, Botswana and Natal, westward to Angola.

Due to the lack of animal food during transport, this is often a very delicate species, particularly the rarely imported race *citerior* from western Africa, the so-called yellow-throated melba finch. Sufficient feeding with animal foods is mandatory. Very aggressive toward its own kind and toward the orange-winged pytilia, but much less so against the aurora finch and the striped waxbill, it should only be kept in pairs together with other grass finches. Feeding: Small grained millet, shelled millet, egg mixture, ant pupae, wax moth larvae, spiders and mealworms (only boiled or freshly molted).

AURORA FINCH, *Pytilia phoenicoptera*

Range: From western Africa to northern Uganda.

This is one of the most frequently imported of all *Pytilia* species and is also one of the hardiest. It is very peaceful, breeds easily and raises the young without difficulties when ant pupae are available in unlimited quantities. Feeding: Small grained millet, half-ripened and germinated spray millet, half-ripened grass seeds, chickweed, ant pupae, egg mixture, wax moth larvae, mealworms and white worms.

The striped waxbill *(Pytilia phoenicoptera lineata)* and the yellow-winged pytilia *(P. hypogrammica)* are close relatives.

DYBOWSKI'S TWIN-SPOT, *Euschistospiza dybowskii*

Range: From Sierra Leone through Nigeria, Cameroon, northeastern Congo to southwestern Sudan.

Habitat: Rocky grass-overgrown areas along river banks and at the outskirts of forests.

Rarely imported, in appearance and vocalization they are similar to the fire finches and especially reminiscent of Jameson's fire finch and the dark fire finch. Feeding: same as fire finches, with lots of fresh ant pupae. Closely related is the dusky twin-spot *(E. cinerovinacea)*.

ROSY TWIN-SPOT, *Hypargos margaritatus*

Range: Northern Zululand, eastern Swaziland and southern Mozambique.

Habitat: Shoreline bushes, acacia thickets and wooded moutain slopes.

The pearl twin-spot is rarely imported, and newly imported birds are extremely delicate until acclimated. Feeding: Canary seeds, lettuce seeds, spray and silver millet, niger seed, ant pupae, half-ripened grass and weed seeds, wax moth caterpillars and white worms.

Yellow-winged pytilia, *Pytilia hypogrammica* (left)—this redder variety is a naturally occurring mutation or hybrid. Orange-winged pytilias, *P. afra* (below).

Opposite: Melba finch, *Pytilia melba* (above left). Aurora finch, *P. phoenicoptera* (above right). Dybowski's twin-spot, *Euchystospiza dybowskii* (below left). Rosy twin-spot, *Hypargos margaritatus* (below right).

PETER'S TWIN-SPOT, *Hypargos niveoguttatus*

Range: From Kenya to southern Mozambique and eastern Angola.

Habitat: Open forests, dense bushes along river banks and brush shady valleys.

Imported in moderate numbers each year, but often delicate while being acclimated, though later on they are hardy and long-lasting. After acclimation and complete molting it breeds in planted aviaries or occasionally in a large flight cage. It is susceptible to disturbances while brooding, and the birds will easily abandon the clutch when the nest is being inspected. Rearing of young is with fresh ant pupae. Feeding: Small grained millet, fresh or refrigerated ant pupae, wax moth larvae, small mealworms, other live insects, spiders, egg mixture, chickweed, grass seeds, lettuce seeds, canary seeds and niger seeds.

GREEN-BACKED TWIN-SPOT, *Mandingoa nitidulus*

Range: From Sierra Leone to southern Ethiopia and southward to northern Angola and Natal.

Habitat: Dense bushes at the edge of the jungle, mountain forests, forest clearings and swampy rice paddies in the proximity of dense bushes. In east Africa it is found at elevations up to 2,000 m.

Rarely imported, delicate for quite some time after arrival and can only be maintained with a constant diet of large amounts of live insects. This bird is considered to be one of the most delicate of the waxbills. Feeding: Small grained millet, half-ripened and germinated spray millet, lots of grass seeds in half-ripened condition, live soft-skinned insects such as fruitfly larvae, flying ants, wax moth larvae and white worms.

BROWN TWIN-SPOT, *Clytospiza monteiri*

Range: From Cameroon to southwestern Sudan and northern Angola.

That various finch species generally get along well with one another is a mainstay of their popularity. Left to right: crimson finch, strawberry, society, chestnut-breasted mannikin, masked grass-finch.

Habitat: In bushes on the savannah and along the edge of forests.

It is very rarely imported, though hardier than the green-backed twin-spot and easier to acclimate. As this bird prefers to stay among dense bushes, it is suited to well-planted aviaries, not cages. Feeding: Same as green-backed twin-spot, and also takes silver and Japanese millet, mealworms and lots of ant pupae. Breeding attempts have so far not been successful.

PARSON FINCH, *Poephila cincta*

Range: Northeastern and eastern Australia.

Habitat: The open, dry savannah, which is characterized by a few intermittent eucalyptus trees.

Since the export restriction for Australian birds, the parson finch has only been available as captive-bred specimens. These progeny are substantially hardier than imported birds and will breed reliably in larger aviaries. They should not be kept together with other *Poephila*

Twin-spots are waxbills native to Africa. This is a male Peter's twin-spot, *Hypargos niveoguttatus.*

Opposite:
Brown twin-spot, *Clytospiza monteiri* (above left). Chubb's green-backed twin-spot, *Mandingoa nitidula chubbi* (above right). Dusky twin-spot, *Euchystospiza cinereovinacea* (below).

The zebra finch, the most common finch, has many color variants: adult gray male and darker billed juvenile (left); fawn, gray and a pair of whites (below).

species as they interbreed easily and hybrids can reproduce. Rearing food: Egg mixture, germinated Senegal millet mixed with cod-liver oil, half-ripened or germinated spray millet, ant pupae, half-ripened grass seeds and seed chickweed. Adults feed on seeds and insects (termites in nature).

ZEBRA FINCH, *Poephila guttata*

Range: Australia, with the exception of northeastern Queensland and the coastal areas of the southeast and southwest of the continent; Lesser Sunda Island (Flores, Sumba, Alor, Timor, Southwest Islands).

Habitat: Open grassland covered with individual bushes and stands of trees close to water.

Apart from the striated mannikin *(Lonchura striata)* and its domesticated form, the Society or Bengalese finch, the zebra finch is the most adaptable and reproductive species of all the grass finches. It is largely domesticated now and has produced not only a series of color mutants, but also has undergone considerable changes in its behavior.

DIAMOND SPARROW, *Zonaeginthus guttata*

Range: Eastern Australia from southern Queensland to eastern South Australia.

Habitat: Dry, open forests with a light tree cover; also lives in close proximity of human and agricultural areas.

Suitable only for aviaries, these birds will easily get too fat in cages and thus breeding results can hardly be expected. Very aggressive toward other grass finches during the breeding season. The family bond breaks down shortly after the young leave the nest, and therefore the juveniles will have to be kept separately. Feeding: Small grained millet, egg mixture, ant pupae, half-ripened grass seeds, germinated and half-ripened spray millet, chickweed, dandelion and sow thistle buds, canary seeds and niger seed.

Since most aviculturists find the parson finch, *Poephila cincta* (left), less attractive than the shafttail, it is not so well established. The different color varieties of zebra finches (below) include (right to left): fawn male, fawn female, white male, gray female, gray male and white female.

The crimson finch, *Poephila phaeton* (right), from Australia is not established in aviaries in the United States. Of the Australian firetails, only the diamond sparrow, *Zonaeginthus guttatus* (below), is established.

GOULDIAN FINCH, *Chloebia gouldiae*

Range: Northern Australia with the exception of the Cape York Peninsula.

Habitat: The dry eucalyptus savannahs in the proximity of rivers and streams.

Newly imported Gouldian finches are excessively delicate and thus have to be kept at elevated temperatures of at least 24°C. A temperature drop below 18°C can be critical, so this bird is unsuitable for aviaries without a heated indoor enclosure. Juveniles are particularly delicate prior to and during the first molt. During this entire period the temperature must be kept constant between 25°C and 30°C. Feeding: canary seed, spray millet, other millets, niger seed, half-ripened and germinated spray millet, half-ripened grass seeds, knot grass and chickweed. Also very fond of leaves and stems of an aloe species *(Aloe arborescens).* Breeds readily, but unfortunately juvenile birds usually die when kept too cold during the molt. Juveniles should not be moved to a new owner until they have acquired the adult plumage.

PAINTED FINCH, *Zonaeginthus picta*

Range: Northern and northwestern Australia, central Australia and western Queensland, primarily in the interior of the continent.

Habitat: Dry, rocky areas (spinifex country) primarily in the proximity of rocky mountain chains. Although it is a ground bird, it is not as closely adapted to the ground as the African quail and grasshopper finches *(Ortygospiza).*

Extremely delicate when newly arrived. Lately, only birds bred in Japan are occasionally imported. Few hobbyists breed this bird with consistent success. Feeding: Small grained millet, shelled millet, half-ripened grass seeds, chickweed, egg mixture and white worms. In nature it feeds exclusively on spinifex seeds.

CRIMSON FINCH, *Poephila phaeton*

Range: Northern Australia, including the Cape York Peninsula, and southern New Guinea.

Habitat: Dry savannahs along rivers and streams, permanently wet coastal areas and often close to human habitation in gardens and parks where open water is available.

Requires lots of heat. Males during the breeding season are extremely aggressive toward all other finches and even toward birds which are substantially larger, such as doves. Therefore this bird should not be kept together with other finches, but instead with larger birds like parakeets. Feeding: Same as other primarily insect-feeding finches (see melba finch).

BLUE-FACED PARROT FINCH, *Erythrura trichroa*

Range: Micronesia, Guadalcanal, New Hebrides, Lifu Island, parts of the Bismarck Archipelago, Cape York Peninsula (Australia), New Guinea, Moluccas and Celebes.

Habitat: Mountain forests, gardens, tree plantations and also on open grassland.

Although rarely imported, hundreds of these birds are being bred each year. It is very shy but active, hardy, durable and easy to breed. To rear the young it is necessary to provide an animal food diet (ant pupae, mealworms, egg mixture) as well as half-ripened spray millet, germinated Senegal millet, canary seed and half-ripened grass seeds. The rather variable diet includes millet, canary seed, shelled and unshelled oats, half-ripened oats and wheat, hemp seeds, weed seeds, ant pupae, mealworms, egg mixture, wax moth larvae, white worms and chicory.

RED-HEADED PARROT FINCH, *Erythrura psittacea*

Range: New Caledonia.

Habitat: Grassland, bushes and neglected plantations.

Hardy, durable and breeding readily, this is one of the

Other Australian firetails are the painted finch, *Zonaeginthus picta* (left), and the beautiful firetail, *Zonaeginthus bellus* (below).

Red-headed parrot finch, *Erythrura psittacea.*

few parrot finches which survived the Second World War in Germany. Ten years after the end of the war there were still some pairs—usually sterile through inbreeding—in the aviaries of some bird fanciers. Pairs were imported from Japan and then from New Caledonia, so nowadays there are adequate stocks available. Feeding: Same as tri-colored parrot finch; will also take fruit (sweet apples, oranges, figs). Preferred rearing foods: Half-ripened grass seeds, half-ripened oats, sow thistle seeds, chickweed, boiled mealworms, ant pupae and knot grass.

PEALE'S PARROT FINCH, *Erythrura cyaneovirens*

Range: Fiji Islands, Samoa.

Habitat: The forest of the interior of the islands, sugar cane plantations, gardens and rice paddies. Rarely imported. More demanding in its diet than the red-eared or blue-faced finch. Feeding: Germinated canary seed, hemp, niger seed, small amounts of millet; prefers half-ripened spray millet, half-ripened rice, grass seeds, knot grass and other weed seeds; also fresh or freshly frozen ant pupae, egg mixture, wax moth caterpillars, apples and figs. Does not breed as easily as the two previously mentioned species.

ROYAL PARROT FINCH, *Erythrura regia*

Range: New Hebrides and Gaua (Banks Islands).

Habitat: Inland forests and mountains up to 2,000 m elevation.

The royal parrot finch is not a geographical race of the short-tailed parrot finch, but instead it is a valid species with different dietary requirements and its own characteristic social behavior. Feeding: Chemically untreated figs (from a health-food store) broken-up so that the birds can get at the flesh and the seeds, boiled mealworms, wax moth larvae, egg mixture, preserved fruit, germinated canary seed, germinated Senegal and spray millet, and niger seed.

RED-EARED PARROT FINCH, *Erythrura coloria*

Range: Mount Katanglad, Mindanao, Philippines.

Habitat: In small clearings in mountain forests, among tall grass and low bushes. Discovered in 1960 by Ripley and Rabor.

This is the smallest Erythrura species. Not as shy as the blue-faced and red-headed parrot finches, but instead adjusts readily to captivity. Feeding: This species requires large amounts of animal protein (fresh or frozen ant pupae, boiled mealworms) and likes to feed on chicory, sweet fruits (apples, pears, oranges) and berries (blueberries).

PINTAILED NONPAREIL, *Erythrura prasina*

Range: Laos, southern Tenasserim to Malaya, Java, Sumatra and Borneo.

Habitat: The edge of forests, bamboo thickets and rice paddies at medium elevations.

Each year large numbers of these birds are imported after being caught in rice paddies. They are being fed during the subsequent transport exclusively on shelled rice. This one-sided and vitamin-deficient diet sustains the birds only for a few weeks and then they invariably die because of severe vitamin deficiency and other metabolic problems. Therefore, it is imperative that they are quickly changed over to germinated wheat and germinated canary seed. Initially, the dried, shelled rice is replaced with rice which has been soaked in frequently replaced water, which is then gradually, and in increasing amounts, mixed with germinated oats until the rice is replaced completely. The germinated oats are then replaced, again gradually, with germinated canary seed. Once the birds accept this diet, germinated oats can be retained as a diet supplement. The nonpareil feeds exclusively on seeds, primarily grass seeds. It is very difficult to breed.

Peale's parrot finch, Erythrura cyanovirens pealii (left). Blue-faced parrot finch, *Erythrura trichroa* (below). Of the parrot finches, breeding stocks of the red-headed and the blue-faced only are established in the United States.

Red-eared parrot finch, *Erythrura coloris* (right). Pin-tailed nonpareil, *Erythrura prasina* (below).

WHYDAHS

PARADISE WHYDAH, *Steganura paradisaea*

Range: From Eritrea and Ethiopia throughout eastern Africa to Zululand and westward to Southwest Africa and Angola.

Habitat: Savannah areas. Brood parasite of the gray-bridled melba finches.

Imported irregularly, usually from Johannesburg, very rarely from East Africa. Quite compatible with all other birds. Only during the breeding season is the male in its nuptial plumage aggressive toward other paradise whydah males with similar plumage. Feeding: Senegal millet, spray millet (both of these also germinated), lots of half-ripened grass seeds, seed-carrying chickweed, lettuce and raw egg yolk. Only birds with intestinal problems will accept egg mixture, ant pupae and wax moth larvae.

BROAD-TAILED PARADISE WHYDAH, *Steganura orientalis*

Range: From Eritrea and Ethiopia to the southern Sudan, westward into the area of Lake Chad. Brood parasite with the East African race of the red-bridled melba finch.

Primary feathers in nuptial plumage in male longer and wider than in the Senegal paradise whydah. The often-applied common name of "yellow-neck whydah" is unsuitable since there are several forms which have a yellow neck band.

WEST AFRICAN PARADISE WHYDAH, *Steganura orientalis aucupum*

Range: From Senegal to the southeastern Lake Chad region. Brood parasite with the yellow-throated melba finch.

Brown nape band. Largest primary feathers of nuptial plumage relatively narrow and short. Most frequently imported species.

GOLDEN-NAPED PARADISE WHYDAH, *Steganura orientalis kadugliensis*

Range: Southwestern Sudan.

Primary feathers of nuptial plumage are somewhat longer, wider and more strongly bent than in *Vidua orientalis orientalis.* Bright yellow neck band. Very rarely imported.

BROAD-TAILED WHYDAH, *Steganura obtusa*

Range: From the eastern and southern Congo to Kenya and southern Rhodesia and to Angola and central Mozambique. Brood parasite with orange-winged pytilia.

Very wide and short primary nuptial plumage feathers. Yellow or yellow-brown neck band. Very rarely imported.

CAMEROONS PARADISE WHYDAH, *Steganura interjecta*

Range: Eastern Cameroon to the Congo, Sudan and, according to the most recent observations, also in Ethiopia. Brood parasite with aurora finch.

Brown neck band. Primary nuptial feathers long and wide. Very rarely imported.

TOGO PARADISE WHYDAH, *Steganura togoensis*

Range: From southern Nigeria to Togo, Ghana and Sierra Leone.

Queen whydah, *Vidua regia.* Though some whydahs have bred in captivity, none are established. Captive breeding of the host species would seem to be a prerequisite.

The paradise whydah, *Steganura paradisea,* is the whydah most frequently imported. In captivity the nuptial plumage often becomes damaged.

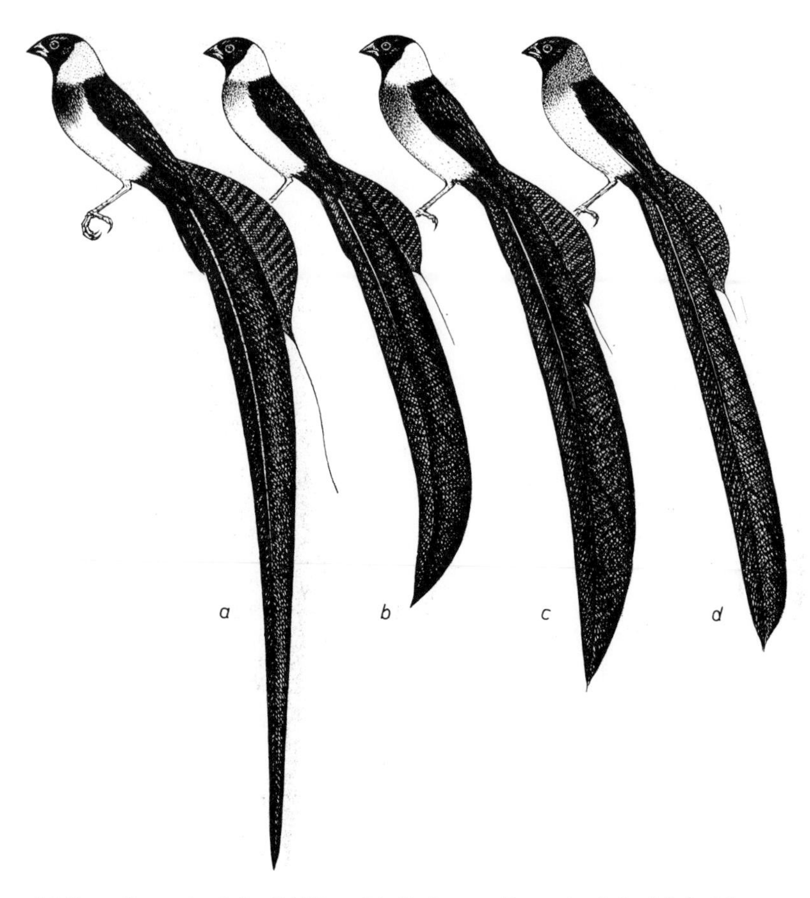

(a) Paradise whydah. *(b)* Broad-tailed paradise whydah. *(c)* Golden-naped paradise whydah. *(d)* West African paradise whydah.

Host bird is the yellow-winged pytilia *(Pytilia hypogrammica).* Brown nape band. The males have the longest and narrowest primary feathers of all paradise whydahs. Often imported together with West African paradise whydahs.

QUEEN WHYDAH, *Tetraenura regia*

Range: Central and southern South Africa, also in a small area in southern Mozambique.

Habitat: Dry thorn-bush steppes. Brood parasite with the violet-eared waxbill.

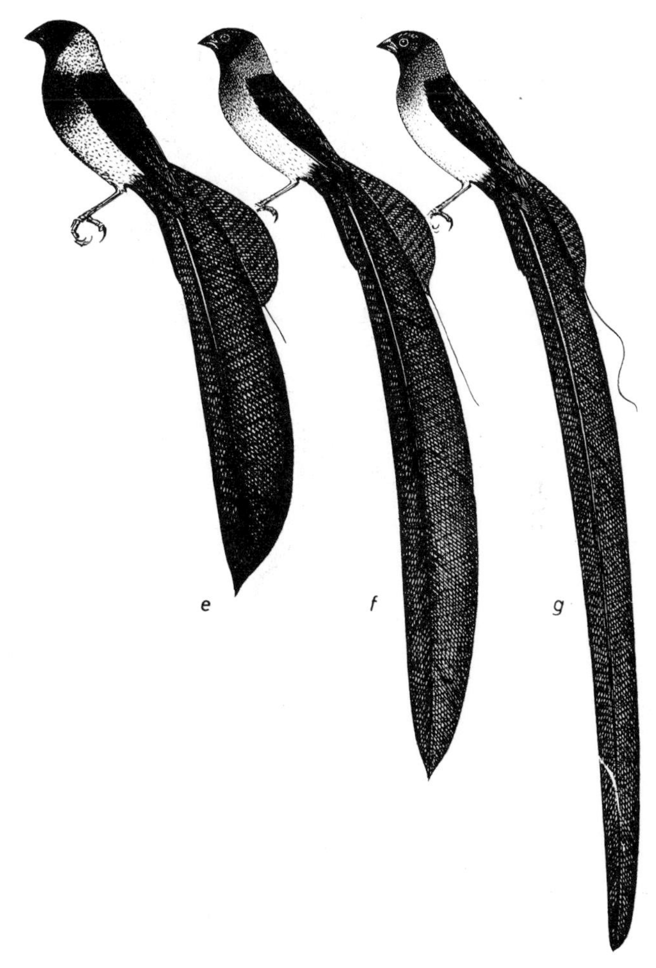

(e) Broad-tailed whydah. *(f)* Cameroons paradise whydah. *(g)* Togo paradise whydah.

Imported together with other South Africa birds (melba finches, violet-eared waxbills, etc.). As is true of all whydahs, it is peaceful and compatible outside the breeding season. After the molt into nuptial plumage the males become exceedingly aggressive against males of their own kind, as well as against other whydah males, particularly Fischer's whydah and the Senegal combassou, and not infrequently also against other bird species. Therefore, this

Male pin-tailed whydahs, *Vidua macroura,* in nuptial plumage.

Opposite:
A pair of pin-tailed whydahs. In eclipse, the male will resemble the female.

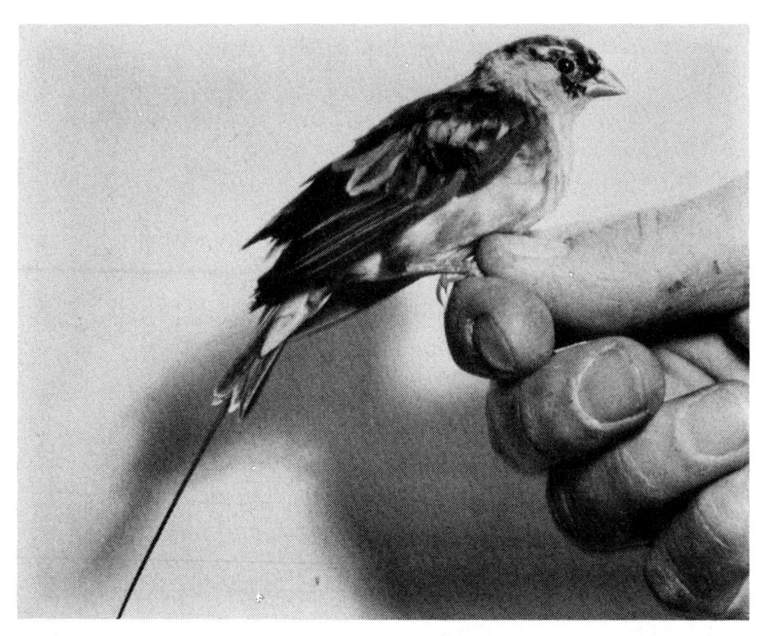

Male queen whydah, *Vidua regia,* molting into eclipse plumage.

species should only be kept in pairs or one male with several females in an aviary. The males copy the vocalization of the violet-eared waxbill. It has not been bred in captivity so far.

FISCHER'S WHYDAH, *Tetraenura fischeri*

Range: From Ethiopia and Somaliland through Kenya to southern Tanzania.

Habitat: The dry thorn-bush steppe. Brood parasite with purple grenadier.

Once rarely imported, in recent times it has become frequently available through the trade. The males are extremely active during the breeding season, constantly vocalize by mimicking the sounds of the purple grenadier and court with incredible persistence. After they have been vocalizing with increasing volume for some time, they begin to beat their wings rhythmically without leaving their position. During this several-minutes-long wing beating the bird orients himself so that he looks directly at the female. After a while he flies in a short arc directly to the female and dances immediately in front of her while strongly beating his wings. If the female is unwilling to mate she will fly away and the male ceases his courtship dance. So far, this species has not been bred in captivity. Females have laid eggs repeatedly in the nests of purple grenadiers kept in the aviary of this author.

STEEL-BLUE WHYDAH, *Vidua hypocherina*

Range: From Ethiopia to southern Tanzania.

Habitat: The dry steppe. Brood parasite with the East African races of the black-cheeked waxbill *(Estrilda erythronotos).*

The resplendent whydah is closely related to the pintailed whydah but not to any of the long-tailed atlas whydahs. Rarely imported, but occasionally available through the trade. Less aggressive than the pintailed whydah, it feeds on eggs. So far, it has not been bred in captivity.

PINTAILED WHYDAH, *Vidua macroura*

Range: Africa south of the Sahara with the exception of jungle areas and deserts; the islands of Fernando Po, Sao Thome, Zanzibar, Mafia.

Habitat: Open grass land with light tree cover. Brood parasite with the red-eared waxbill and possibly other *Estrilda* species.

Imported regularly. Hardy and long-lived. It is very aggressive toward other whydah males in nuptial plumage during the breeding season and, therefore, should not be kept together with other whydahs, although it is not as vicious as the queen whydah. Feeds on eggs, as do all whydahs (will even attack large dove eggs), so therefore should not be kept in breeding aviaries. Only the eggs of

The Senegal combassou, ***Hypochera chalybeata*** (left), is the only member of the parasitic whydahs in which the male does not have the characteristic elaborate plumage. Other groups in this family (Ploceidae) include the bishops, or weavers: Napoleon weaver, *Euplectes afra* (below), and orange weaver, *Euplectes orix franciscana* (opposite above); and such sparrows as the golden sparrow, *Passer luteus* (opposite below).

(1) Queen whydah. *(2)* Fischer's whydah. *(3)* Steel-blue whydah. *(4)* Pintailed whydah. *(5)* Senegal combassou (race: *Hypochera c. chalybeata).*

parakeets in deep nest boxes are safe from these birds. Breeding has been successful several times with red-eared and other waxbills as host birds. Even society finches can raise young pintailed whydahs when these can be persuaded to use ant pupae and egg mixture as rearing foods as long as possible; the foster pair must not have access during the rearing period to large grain millets (silver millet, Morocco millet and others). Apart from the vital animal proteins, only Senegal and spray millets (half-ripened or germinated) should be fed. Feeding: Small grained millet, half-ripened grass seeds, half-ripened or germinated spray millet, germinated millet mixed with cod-liver oil, seed-carrying chickweed, lettuce and raw chicken eggs.

SENEGAL COMBASSOU, *Hypochera chalybeata*

Range: From Senegal to Ethiopia and Eritrea in three races *(chalybeata, neumanni, ultramarina)*, an additional one *(centralis)* in parts of eastern Africa and a red-beaked race *(amauropteryx)* in southwestern and eastern Africa southward to eastern Zululand. A brood parasite of the red-billed fire finch *(Lagonosticta senegala)*.

The western races are frequently imported, the eastern ones very often, but the red-beaked race rarely appears on the market. There is a bluish sheen to the plumage. All are hardy and long-lived. While in nuptial plumage the males are peaceful with all other birds, but not with other whydah males. The vocalization imitates that of the fire finch, including the begging sounds of juveniles. It has been successfully bred.

Mutations that probably would not survive under wild conditions are perpetuated in captive breeding programs. This applies, for example, to canary varieties. North Hollander frill (left). Pair of border canaries (below). Gloster corona hen with her chicks (opposite above). Pair of lizard canaries (opposite below).

Finches

CANARY, *Serinus canaria*

Range: Azores, Madeira and western Canary Islands. Widely introduced.

Habitat: Open forests, fruit orchards, gardens and tree stands along roads.

This species gave rise to all domesticated canaries. The red so-called "color" canaries are genetically not pure canaries but are hybrids between the hooded siskin and canaries. Wild-caught canaries are rather delicate; they do not do well in captivity and are very difficult to breed. They suffer frequently from mite infections of the feet.

CITRIL FINCH, *Serinus citrinella*

Range: Mountains of central and southern Europe.

Habitat: Open evergreen and mixed forests in mountain chains, edges of forests, open meadows and tree stands.

The most difficult of all the native finches. Susceptible to high summer temperatures and difficult to breed. A rather versatile seed eater: Mixed feed with lots of niger seed, lettuce seed, pine seeds, poppy seeds, half-ripened seeds (dandelion, sow thistle, other garden weeds), ant pupae and aphids.

SERIN, *Serinus serinus*

Range: Western north Africa, eastward to Libya, southern and central Europe to Asia Minor and towards the western Ukraine, northward to the Baltic countries and northern Germany.

A rather adaptable grain-feeder that is hardy and long-lived and breeds easily in sufficiently large aviaries. The males will easily hybridize with canary females; male hybrids are fertile, but female hybrids are sterile. Feeding: Mixed seed with plenty of summer rape seeds and half-ripened weeds seeds.

CAPE CANARY, *Serinus canicollis*

Range: From eastern Africa (Ethiopia) to southern and eastern South Africa and Angola; introduced into Mauritius and Reunion.

Habitat: Mountains and highlands.

Imported irregularly, it is hardy but difficult to breed. Hybrids with canaries are often produced; fertility of these hybrids is unknown. Feeding: Same as the citril finch, which is closely related to this species.

ALARIO FINCH, *Alario alario*

Range: South Africa (except Natal and the northern and eastern Transvaal) and Southwest Africa.

Despite strongly deviating coloration, this is a close relative of the cape canary. Although somewhat delicate when newly imported, it is durable after being well-acclimated. Breeding is difficult, and the species tends to hybridize with canaries. Feeding: Same as citril finch; lots of niger seed, Senegal millet, lettuce seed, dandelion, sow thistle and grass seeds.

YELLOW-RUMPED SERIN, *Ochrospiza atrogularis*

Range: Southern and southwestern Africa northward to Angola and through the eastern Congo and Western Tanganyika to southern Uganda. The recently often imported Eritrean serin *(O. xanthopygia)* from northern Africa is not a geographical race of the Angolan serin, but is a valid species.

The "true finches" (Fringillidae) include the alario finch, *Serinus alario* (opposite above); the citril finch, *Serinus citrinella* (opposite below); and the Himalayan siskin, *Carduelis spinoides* (above).

Infrequently imported. Strongly recommended as an aviary bird, it is hardy and long-lived, breeds easily (even in a large cage) and has a full, variable song. The young should be raised with germinated millet, heads of lettuce and various green foods (chickweed). Feeding: Senegal and spray millet, niger seed, poppy and lots of green feed.

ST. HELENA SEEDEATER, *Crithagra flaviventris*

Range: Southern Angola and southwest Africa to South Africa (except in the largest section of Natal and eastern Transvaal); introduced into St. Helena.

Imported irregularly, it is hardy and long-lived. It breeds easily and raises its young reliably on egg mixture, ant pupae and half-ripened dandelion and sow thistle seeds. Hybrids with canaries are sterile. Feeding: Lots of niger seed, shelled oats, lettuce seeds, linseed, canary seed, green food and ant pupae.

EURASIAN SISKIN, *Carduelis spinus*

Range: From Iceland to Japan, northward through the north Scandanavian countries and southward to northern Iran; largely absent from central Siberia. Habitat: Dense evergreen forests; medium to high mountain chains up to the tree line.

Easy to acclimate and the hardiest of all siskin species. Juvenile birds become easily hand-tame and will accept hemp and other seeds if they are offered directly. Breeds easily. Feeding: Mixed seeds with lots of niger seed, lettuce seed and birch, alder and pine seeds. The main food items, niger seed and poppy, should be offered in separate dishes in addition to the mixed seed. Also eats lots of half-ripened seeds (sow thistle, dandelion, lettuce, mugwort, sorrel, etc.), egg mixture and fresh ant pupae. Feeds in nature on a variety of seeds such as birch and alder during the breeding season, also caterpillars and aphids.

BLACK-HEADED SISKIN, *Carduelis magellanicus*

Range: A vast area of South America, from the Colombian Andes to Chile, northern Patagonia, eastern Venezuela, Guyana and eastern Brazil from Bahia southward to Uruguay and eastern Argentina.

Imported regularly, mostly males. They are often sick when they have been fed exclusively with canary seed and millet before reaching the pet shop. The species has been bred several times, but clutches are often unfertilized. The juveniles undergo a complete molt at the age of three months. Feeding: Niger seed (must always be available in unlimited quantities), poppy, lettuce and thistle seeds, some canary seed, lots of half-ripened grass and similar seeds, seed-carrying chickweed, egg mixture and fresh ant pupae. This bird is extremely susceptible to damp and cold; it must be kept at about 22^0 to 24^{0}C during the winter.

LESSER GOLDFINCH, *Carduelis psaltria*

Range: Western United States throughout Central America to Colombia, Ecuador, Peru.

Much hardier and longer lived than the black-hooded siskin and the Yarrell's siskin, it breeds easily in spacious outdoor aviaries which are planted with bushes or small trees. The small nest is constructed among dense bushes from coconut fibers, dried grass and cotton threads. Usually it accepts artificial nesting facilities (baskets). To raise the young there should be an ample supply of half-ripened dandelion, chickweed and sow thistle seeds. Feeding: Accepts almost exclusively niger seed, plus some lettuce seed, poppy, thistle and canary seed, egg mixture and fresh ant pupae. It has to be kept in a well-heated room at temperatures from 22^0 to 24^{0}C during the winter months.

The serin, *Serinus serinus,* a common resident of southern Europe, widely kept by European aviculturists.

The Eurasian siskin, *Carduelis spinus,* a frequent candidate for cross-breeding with canaries.

BLACK SISKIN, *Carduelis atratus*

Range: Northern Chile, western Argentina, Bolivia and southern Peru. Occurs up to elevations of 2,600 m.

Bird fanciers are strongly advised against acquiring and keeping this bird. So far, the newly imported birds have always died within a few weeks despite a satisfactory and correct diet. Therefore, this species can apparently not be maintained in the temperate climates with the food items available there.

YARRELL'S SISKIN, *Carduelis yarrelli*

Range: Eastern Brazil (southern Bahia to Ceara) to northern Venezuela.

Rarely imported and very delicate. Most specimens available are sick and/or diseased due to an insufficient diet (little or no niger seed in the diet (and are susceptible to cold and dampness. Must be kept very warm. So far it has not been bred. Feeding: Same as lesser goldfinch.

HOODED SISKIN, *Carduelis cucullatus*

Range: Venezuela and Trinidad.

This fire-red bird is one of the most highly desired tropical finches, but unfortunately far too many of these birds are purchased by inexperienced hobbyists, usually canary breeders who use them for hybridization. Most of them do not survive very long on canary seeds. Since the populations in the wild are becoming strongly endangered due to excessive collecting, hobbyists should make considerable efforts to breed this bird in captivity. Only in this way can the limited stocks be increased and the bird trade become independent of imported animals. It breeds relatively easy even in larger cages, where this bird readily accepts canary nests. Rearing and Feeding: Same as lesser goldfinch. Must be kept at temperatures of 22° to 24°C during the winter months.

The linnet, *Acanthis cannabina,* often kept because of its sweet song.

LINNET, *Acanthis cannabina*

Range: From western North Africa, Canary Islands and Madeira via Spain throughout all Europe to western Siberia and Asia Minor.

Habitat: Bush- and tree-covered open areas, cemeteries, small tree stands and large agricultural areas adjacent to human habitation.

Newly collected birds are invariably very shy and frighten easily, so they are more suitable for aviaries than for cages. They are hardy and long-lived. The birds breed easily and pair readily with canaries. Young males often copy the vocalization of canaries when they have been kept together with these birds in aviaries, but, gradually this is replaced by their own species-specific vocalization. This bird also likes to take a fair amount of summer rape seed.

There is not enough evidence to support the claims of some aviculturists that the twite, *Acanthis flavirostris,* is a quarrelsome bird.

Opposite:
Chaffinch, *Fringilla coelebs* (above).
Black-headed siskin, *Carduelis magellanicus* (below).

TWITE, *Acanthis flavirostris.*

Range: Northern England to Scandanavia; also Asia Minor to Mongolia.

Habitat: Steppes and steppe-like tundras, rocky grassland, heath regions and high plateaus.

Rarely available, but very hardy and long-lived. Hybridizes easily with canaries, common redpolls, and linnets, but the fertility of these hybrids is still unknown. Feeding: Same as Eurasian linnet.

PINE GROSBEAK, *Pinicola enucleator*

Range: Circumpolar in forest zones and in mountains.

Habitat: Open evergreen forests.

The pine grosbeak is a seldom-collected migratory bird. It is very susceptible to heat and, therefore, it should only be kept in cool, shady outside aviaries; it must also spend the winter there. Strong solar radiation during the summer and intense indoor heating during the winter are difficult for the birds to cope with, and most of them do not survive the molt. Feeding: Various berries, sunflower seeds, hemp, rape, linseed, canary seed, shelled oats, pine seeds, apples and pears, tree buds, pine shoots, green feed, ant pupae, egg mixture, mealworms and other live insects. Breeds easily. During the first week of their life the nestlings are fed exclusively on insect food.

BULLFINCH, *Pyrrhula pyrrhula*

Range: From western Europe (northern Spain) to Japan. In the Lake Baikal region there occurs a gray-breasted form *(cineracea)* and in Japan one form with a gray abdomen, red throat and red cheeks *(griseiventris).* The bullfinch *(Pyrrhula murina)* which used to live on the Azore island of St. Miguel is presumably extinct.

Habitat: Pine tree forests of the plains and mountains. In western Europe it also inhabits mixed forests, gardens, parks and cemeteries with conifer stands.

Fifteen-day-old bullfinch juveniles.

Bullfinches form permanent pair bonds and thus they should be kept only in pairs. Only hand-raised birds which have become imprinted upon humans and which are strong "singers" do not need the company of their siblings if they have been kept individually longer than one year. Accommodations should be large, 80 to 100 cm-long cages or, even better yet, outdoor aviaries planted with various bushes with tough leaves. Pine trees are invariably destroyed by bullfinches within a very short period of time, since they bite off all new shoots. Feeding: A mixture of niger seed, canary seed, rape seed, shelled oats, linseed, various pine seeds, evening primrose seed, hemp, lettuce seeds, thistle seeds, sunflower seeds and dried mountain ash berries. These birds should also receive carrots, apples, chicory and lettuce, as well as half-ripened chickweed seeds, dandelion, sow thistle, knot grass, shepherd's purse and pansies. Fresh ant pupae are eagerly accepted. In order

The bullfinch, *Pyrrhula pyrrhula,* cannot be said to be well established in American aviculture. Perhaps fanciers find this bird's special dietary requirements too difficult to meet.

One of the most popular cagebirds in Europe, the goldfinch, *Carduelis carduelis,* has not been widely kept in the United States.

to raise the young, half-ripened seeds, especially those of dandelion and sow thistle, have to be available in large amounts.

GOLDFINCH, *Carduelis carduelis*

Range: North Africa and Europe to Mongolia and the western Himalayas.

Habitat: Open, tree-covered areas, gardens, orchards and tree stands along roads. Versatile seed-eater, especially those of composite flowers or plants.

The most frequently kept European finch. Losses at most dealers are still fairly high, since these birds are usually fed exclusively on a mixed diet which is low in the required types of seeds (niger seed, lettuce seeds and poppy). Most of the components in these mixed diets are not taken very willingly by these birds (linseed, shelled oats, canary seeds) or are only taken when extremely hungry (rape seed, millet). Since a large amount of these seeds invariably leads to severe digestive disturbances, young birds especially succumb easily under such a diet. Therefore, niger seed and poppy seed have to be offered in separate containers so that these are available to the birds in unlimited quantities. This also applies when goldfinches are kept as individual birds. The goldfinch diet must contain large amounts of niger seed, lettuce seed, poppy, thistle seed, mountain seeds, pine seeds and hemp, as well as some canary seed and shelled oats. Rape seed and millet are unsuitable. During spring, summer and fall, substantial amounts of half-ripened dandelion and sow thistle seeds, as well as chickweed and knot grass, should be offered.

Thrushes

ROCK THRUSH, *Monticola saxatilis*

Range: From Spain and western North Africa throughout the Mediterranean area and to northern China.

Habitat: Sunny slopes, rocky plains with large boulders, quarries and ruins; a mountain bird.

For some years now, especially the fall months, this bird has been frequently available through the European trade. Most of the rock thrushes available are hand-raised juveniles from Italy. They are most suitably kept in a spacious aviary which is equipped with rocks and bricks (attached halfway up the wire and in the corners); cage birds must have gypsum- or cement-coated perches and similarly treated facilities. Feeding: Ant pupae, soft food, egg, mealworms, large insects such as crickets and grasshoppers, wax moth caterpillars, etc. Furthermore, the birds should get, throughout the year, grapes, juniper berries, chopped sweet apple, etc. Rock thrushes reach an age of up to 18 years and are bred frequently.

BLUE ROCK THRUSH, *Monticola solitarius*

Range: From Spain and western North Africa through the western and northern regions of the Mediterranean to eastern China, southeastern Siberia, Taiwan, Malaya and Japan.

Habitat: Occurs in drier, warmer habitats than the rock thrush and, therefore, is not found at high elevations.

Old world thrushes include the shama thrush, *Copsychus malabaricus indicus* (opposite); the bluethroat, *Luscinia svecica* (right); and the robin, *Erithacus rubecula* (below).

The shama's ability as a mimic is the basis for its popularity among fanciers.

Also imported throughout the year from Italy as juveniles and as wild-caught birds. Hand-raised birds become very tame, sit on your hand and court the keeper. In suitable areas they can be easily conditioned for free-flying; however, at night they should be baited back into the aviary and kept there until the next day. Blue rock thrushes kill half-grown mice and can be dangerous to incapacitated small birds. Apparently this bird has not yet been bred.

SHAMA THRUSH, *Copsychus malabaricus*

Range: India, Sri Lanka, Pakistan, Burma, Thailand, Indo-China, Malaya, Andamans, Sumatra, Java and Borneo.

Habitat: Tropical rain forest, bamboo jungles in the proximity of rivers and teakwood forests. Many geographical races; those living on northern Borneo *(stricklandi* and *barbouri)* have a white crown.

Shama thrushes are frequently imported from India as juveniles, which are easy to capture. They are superb singers with a highly variable repertoire, constantly combining new phrases with already established ones. They imitate many other birds. Because of their long tail, the cage has to be very spacious (80 to 100 cm long). They are very hardy and long-lived (up to 26 years) when given the proper diet. Feeding: Fresh ant pupae, soft food with cottage cheese, egg mixture, live insects such as grasshoppers, crickets, wax moth larvae, centipedes and chironomid midge larvae; also takes spiders and small earthworms. Berries (juniper) are only taken occasionally, but they will accept rather eagerly finely cut green feed (chickweed, lettuce), pine cones and cashew nuts. Can easily be trained for free-flying provided there are dense bushes close to the building. Breeds easily and raises the young with live insects and fresh ant pupae. The shama thrush is a cave or semi-cave brooder.

SLATE-COLORED SOLITAIRE, *Myadestes unicolor*

Range: Mexico.

Rarely imported from Mexico, usually together with the mountain, or brown-backed, solitaire, *M. obscurus,* which also occurs there. An excellent singer with a repertoire of about five to six different verses which are usually repeated several times. Individual tones sound like musical instruments, are strangely quavering and are comparable to other bird sounds. The clarino and its relatives are unsuitable for continous cage maintenance. They require a spacious aviary with dried branches (not too many leaves). Feeding: Berries and fruit (finely chopped sweet apple) are the main diet. Among those berries preferred are juniper, elderberries, bilberries, blackberries and cranberries, as well as grapes and raisins soaked in water. Soft food is not eagerly taken; however, the birds are very fond of fresh and dried ant pupae, soft-skinned insects, wax moth caterpillars, freshly molted crickets, egg mixture and honey water. So far, this bird has been bred only a few times.

The chestnut-bellied rock thrush, *Monticola rufiventris,* an Asian member of the thrush family.

Opposite:
Two European members of the thrush family, the nightingale, *Luscinia megarhynchus* (above), and the thrush nightingale, *Luscinia luscinia* (below), are similar in appearance and song, but they do not occupy the same range. Few birds have been mentioned in literature as often as the nightingale.

GULDENSTADT'S REDSTART, *Phoenicurus erythrogaster*

Range: From the Caucasus and northern Iran to the mountains of Mongolia.

Habitat: Mountains, rocky high plateaus, alpine tundras and steppes, often in the proximity of rapidly flowing rivers. A non-migratory bird which only moves during the winter months to lower elevations. Feeding: Same as shama thrush.

SIBERIAN RUBYTHROAT, *Luscinia calliope.*

Range: From the Ural Mountains to easternmost Siberia, northern Japan and western China.

Habitat: Dark, swampy pine forests.

An Indian import, where this species spends the winter, it is substantially hardier and longer-lived than the related bluethroat. It is very suitable for aviaries since it is very compatible with other small birds though very aggressive toward its siblings and other ground singers. The aviary should be planted with small pine trees and other conifers with branches reaching down to the ground. The ground should be thickly covered with pine needles, which have to be sprayed with water daily during prolonged periods of drought. The males often sing during warm summer rain showers. Prefers to take chironomid midge larvae and similar insects, as well as white worms and very small earthworms. Very difficult to breed.

EASTERN RUBYTHROAT, *Erithacus pectoralis*

Range: From eastern Afghanistan through Turkestan and the Himalayan region to China and northern Burma.

Habitat: Breeds in dense mountain forests, even above the tree line at elevations of 3,000 to 4,000 m.

Rarely imported. Maintenance and feeding are the same as for the Siberian rubythroat, but this species prefers a drier, partially rocky aviary bottom. The aviary should be planted with rhododendron, dwarf juniper and pine trees. So far, this species has been bred only once.

ROBIN, *Erithacus rubecula*

Range: Atlantic Islands (Madeira, Canary, Azores), western North Africa and Europe to western Siberia, Asia Minor and northern Iran.

Habitat: Deciduous and mixed forests, gardens, parks, cemeteries, etc.

A non-demanding insect-feeder which is easy to acclimate and to keep. It is unsuitable for keeping in aviaries with other small birds, since this bird can become extremely aggressive after having behaved quite peacefully for many months. Many robins cease singing after a winter in captivity. The reason for this is incomplete gonadal growth caused by disturbances in the seasonal rhythm (too much light during the winter months). Feeding: Fresh ant pupae, soft food, egg, meal worms and other insects, spiders, various berries and cherries. It is often bred in captivity.

BLUETHROAT, *Luscinia svecica*

Range: From western Europe throughout northern and central Asia to northeastern Siberia and western Alaska, southward to northern Iran and to the northwestern Himalayas.

Habitat: Swampy areas with alder-tree stands as well as river and lake banks with extensive reed and willow tree stands, also ditches in agricultural regions and periodically even along wet slopes.

The bluethroat robin is highly endangered throughout central Europe, so that its collection and captive maintenance can no longer be justified. Therefore, the following details are intended to provide instructions to maintain injured birds. Feeding: Fresh or freshly frozen ant pupae, soft food with lots of white worms, egg mixture,

The thrushes are a large family that range throughout the world. The most familiar member of the family in the United States is the American robin. The birds shown on this and the following page have a European range. This is a fledgling mistle thrush, *Turdus viscivorus*.

Opposite:
A wheatear, *Oenanthe oenanthe* (above), and an adult mistle thrush (below).

caddis fly larvae, chironomid midge larvae, daphnia, juniper berries and other berries. The bottom of the cage has to be covered with damp moss (which must be changed daily), and it should contain a large shallow container for bathing.

NIGHTINGALE, *Luscinia megarhynchos*

Range: Western North Africa, southwestern and central Europe, Asia Minor and Iran to northern Afghanistan.

Habitat: Dry deciduous and mixed forests, parks, gardens and cemeteries. Prefers a drier area than the thrush nightingale.

Adapts easily to captivity, but it is susceptible to sudden position changes of the cage. If the cage is being changed around during the singing period, most birds will cease to sing. Feeding: During the summer months they prefer exclusively fresh ant pupae, live insects and egg mixture. Mealworms in large quantities (10 to 12 daily) should only be given during the singing period; outside this period two or three mealworms at the most, should be given daily—otherwise swellings will develop on the feet of these birds. Relatively easy to breed in a well-planted aviary.

THRUSH NIGHTINGALE, *Luscinia luscinia*

Range: From central Europe to Asia Minor and China.

Habitat: Wetter areas than the Eurasian nightingale.

This bird is somewhat less delicate than the Eurasian nightingale. It has been bred repeatedly in well-equipped and planted aviaries.

Warblers and Flycatchers

GARDEN WARBLER, *Sylvia borin*

Range: From Portugal, central Spain and the British Isles in the west to western Siberia in the east.

Habitat: Mixed forests with dense underbrush, gardens and parks.

This bird is far more difficult to keep and far more delicate during the winter molt than the monk warbler. Feeding: Must have fresh ant pupae during the summer months; during fall and winter it needs a not-too-rich soft food. It should receive lots of fruit such as apple, pieces of pear, plums, cherries, peaches, apricots, oranges, grapefruit and a variety of berries.

BLACKCAP WARBLER, *Sylvia atricapilla*

Range: From the Azores, Portugal and North Africa in the west and south to western Siberia.

Habitat: Similar to the garden warbler, but also occurs in pine forests.

The easiest to keep of all warblers, which should not encourage low-grade food and diet. It molts prior to moving into winter quarters. Feeding: Same as garden warbler, with lots of berries and fruit, including items such as bananas, dates, figs and raisins. However, the latter should only be offered with caution since these fruits have a tendency to fatten the birds. Has been bred frequently.

The blackcap, *Sylvia atricapilla;* the bird shown is a female.

Opposite:
The garden warbler, *Sylvia borin* (above), and the blackcap are European counterparts of the American wood warblers. The satin flycatcher, *Myiagra cyanoleuca* (below), illustrates the interesting coloration typical of Australian flycatchers; the male of this dimorphic species is shown.

LITTLE BLUE-AND-WHITE FLYCATCHER, *Muscicapula superciliaris*

Range: From the northwestern Himalayan region to western China and Burma.

Habitat: Deciduous, pine and mixed forests.

Very rarely imported. A very demanding and very active insect-feeder which should only be kept by experienced aviculturists. Feeding: Fresh ant pupae, high quality soft food and lots of small live insects. Has not been bred so far.

TICKELL'S BLUE FLYCATCHER, *Cyornis tickelliae*

Range: India, Sri Lanka, eastern Pakistan, Burma, Malaya, Indo-China, Philippines and Celebes.

Habitat: Rain forests, bamboo thickets and plantations.

Irregularly imported, but in recent years available more frequently. A very adaptable bird that is a good singer with a song which is more variable and harmonious than that of the red-bellied niltava and resembles that of the Indian white-eye *(Zosterops palpebrosa).* An excellent aviary bird, but it is often aggressive toward other flycatchers. Likes to take finely chopped sweet apples and pears, chopped cashew nuts and pine seeds, flies, daphnia and chironomid larvae.

RUFOUS NILTAVA, *Niltava sundara*

Range: Himalayan region to Burma and Indo-China.

Habitat: Wet evergreen forests, occasionally also pine forests with dense undergrowth.

Imported infrequently from India, the magnificently colored niltava catches flying insects but also often goes to the ground to pick up prey there. It is less specialized than the verditer or blue-and-white flycatchers and an excellent aviary bird, though less suitable for cage maintenance. It is compatible with all other birds but aggressive toward other flycatchers. Feeding: Ant pupae, soft food with cottage cheese, egg mixture, lots of live insects and small portion of berries. Has been bred repeatedly in Germany and England.

VERDITER FLYCATCHER, *Eumyias thalassina*

Range: Himalayan region to western China and Malaya, eastern Pakistan and Burma.

Habitat: Deciduous forests, especially open oak forests, and also pine forests.

Infrequently available, it is usually imported from India. A conspicuously short-legged, long-winged flycatcher, it is unsuitable for permanent cage maintenance. In the aviary it is compatible toward all birds except other flycatchers, sometimes leading to antagonistic behavior. Feeding: Same as the niltava, but more demanding; requires even more live or freshly frozen insects (wasp pupae, moths).

BLUE-AND-WHITE FLYCATCHER, *Cyanoptila cyanomelana*

Range: from the Kuril Islands and Manchuria to northeastern China, Japan and Korea.

Rarely imported. It is short-legged and long-winged and, therefore, a better flycatcher than the niltava, though less suitable for cage maintenance. Very active in the aviary, it spends most of the day catching flying insects and touches the ground rarely, and then only to drink, to bathe and to pick up food insects thrown into the aviary. In large aviaries these birds will become tame quickly and will take wax moth caterpillars and other insects from the hand. Feeding: Same as the niltava, but somewhat more demanding in its food item selection and takes more berries. Has apparently not yet been bred; females are never or only rarely imported.

The spotted flycatcher, *Muscicapa striata,* is the most wide-ranging European flycatcher.

Like the satin flycatcher, the black-faced flycatcher, *Monarcha melanopsis,* is one of the monarch flycatchers. The monarchs differ in behavior from typical flycatchers in that they feed more extensively in trees and on the ground.

Doves

DIAMOND DOVE, *Geopelia cuneata*

Range and habitat: Dry areas of Australia; not in the coastal region.

Bred each year in large quantities, this dove is very compatible with all other birds. During the breeding season the males enter into mild, harmless fights with other doves. Several pairs can be kept in large aviaries, provided they are all introduced at the same time. The males spar frequently with each other, but without disturbing the breeding activities of the other pairs. In small aviaries one should keep only one pair together with other small birds (finches, insectivores, quail). It requires lots of heat and direct sun and, therefore, in the long run these birds must be kept in aviaries exposed to lots of sunshine, but with an adjacent protected, heated enclosed space. The incubation period is thirteen days; the young will leave the nest after ten to twelve days. Females are sexually mature after three months (the earliest recorded onset of egg-laying is at the age of 77 days), while males reach sexual maturity after three and a half to four months. The most favored nesting material is coconut fibers of 10 to 12 cm in length. The small basket-like nest is built free-standing in bushes or pine trees.

Diamond doves do not bathe, but instead like to rest on fresh, cool soil or cool or only slightly moist chickweed or grass. The primary foods are Senegal millet and poppy. In

Because of their habits, ground doves are more suited to aviaries than cages. Blue ground dove, *Claravis pretiosa* (right). Pygmy ground dove, *Columbina minuta* (below).

One reason why diamond doves, *Geopelia cuneata,* are widely kept is their compatibility with other bird species.

Opposite:
There are many similarities between Old- and New-World quail doves. The blue-headed quail dove, *Starnoenas cyanocephala* (above), is from Cuba, while the golden-heart dove, *Gallicollumba rufigula* (below), is native to New Guinea.

addition, these birds also occasionally take some of the seeds included in wild bird mixtures. Green food (chickweed, lettuce) and calcium-rich substances should be constantly available. Diamond doves can be used as foster parents for other small dove species, including ruddy, Peruvian, and pygmy ground doves and Galapagos doves.

PYGMY GROUND DOVE, *Columbina minuta*

Range: Southeastern Mexico, all of Central America and locally in South America to Paraguay and the coastal region of Peru.

Apart from the diamond dove, this is the smallest of the dove species. It should not be confused with the passerine ground dove *(Columbina passerina)*, which is colored differently and has an orange beak with a black tip (the beak of the pygmy ground dove is completely black). It breeds readily but is easily disturbed by other birds, particularly when kept in small aviaries or flight cages. Under these conditions, parents readily abandon the clutch or the nestlings. Nesting requirements and feeding: Same as for the diamond dove.

RUDDY GROUND DOVE, *Columbina talpacoti*

Range: From Mexico throughout Central America to eastern Peru, Argentina and southern Brazil.

Hardy, undemanding and easy to breed, it requires warmth and direct sunshine and very unsuitable for indoor aviaries or cages. The plumage of birds kept out of the sun gradually becomes black and the beak and claws grow excessively long and have to be constantly trimmed. It is quite compatible with other birds, but during the breeding season there may be some fighting with other doves of similar size. It likes to bathe and sometimes swims over short distances. During the winter it must be kept warm, the same as the other dove species discussed here (from October to the middle of April). Feeding, nesting facilities and nesting material are the same as those of the diamond dove.

PERUVIAN GROUND DOVE, *Columbina cruziana*

Range: Northern Ecuador to northern Chile.

Habitat: Dry coastal regions.

Distinguished from other small South American ground doves by the presence of a violet-brown band over the shoulder, the males have an orange beak with a black tip and the females have a yellow beak. The only sound is a grunting-humming noise emitted by the male, especially during the courtship. It is very aggressive toward other small doves but breeds easily and reliably, raising five to six broods successively.

BLUE GROUND DOVE, *Claravis pretiosa*

Range: From southern Mexico to Peru, Bolivia, northern Argentina and southern Brazil.

Rarely imported because this bird is not abundant in its natural range, the male is delicately blue-gray and has some large bluish black spots on its wings. Females are cinnamon-brown with dark brown wing coverts. This bird is identical in its shape and behavior as the rusty-colored ground dove but is very much larger. It breeds easily and reliably and is compatible and long-lasting. Feeding: Same as other small doves (see diamond dove), but will also take larger millets (silver millet).

In the same genus *(Geopelia)* as the diamond dove, the zebra dove, *G. striata striata,* has similar requirements.

Opposite:
The Peruvian ground dove, *Columbina cruziana* (above), and the ruddy ground dove, *Columbina talpacoti* (below), are among the most frequently kept species of American ground doves.

Index

Page numbers in parentheses refer to illustrations.

Bar-shouldered dove, *Geopelia humeralis.*

Opposite:
Spotted turtle dove, *Streptopelia chinensis.*